IB learner profile

The aim of all IB programmes is to develop internationally minded people who, recognizing their common humanity and shared guardianship of the planet, help to create a better and more peaceful world.

IB learners strive to be:

Inquirers They develop their natural curiosity. They acquire the skills necessary to conduct inquiry and research and show independence in learning. They actively enjoy learning and this love of learning will be sustained throughout their lives.

Knowledgeable They explore concepts, ideas and issues that have local and global significance. In so doing, they acquire in-depth knowledge and develop understanding across a broad and balanced range of disciplines.

Thinkers They exercise initiative in applying thinking skills critically and creatively to recognize and approach complex problems, and make reasoned, ethical decisions.

Communicators They understand and express ideas and information confidently and creatively in more than one language and in a variety of modes of communication. They work effectively and willingly in collaboration with others.

Principled They act with integrity and honesty, with a strong sense of fairness, justice and respect for the dignity of the individual, groups and communities. They take responsibility for their own actions and the consequences that accompany them.

Open-minded They understand and appreciate their own cultures and personal histories, and are open to the perspectives, values and traditions of other individuals and communities. They are accustomed to seeking and evaluating a range of points of view, and are willing to grow from the experience.

Caring They show empathy, compassion and respect towards the needs and feelings of others. They have a personal commitment to service, and act to make a positive difference to the lives of others and to the environment.

Risk-takers They approach unfamiliar situations and uncertainty with courage and forethought, and have the independence of spirit to explore new roles, ideas and strategies. They are brave and articulate in defending their beliefs.

Balanced They understand the importance of intellectual, physical and emotional balance to achieve personal well-being for themselves and others.

Reflective They give thoughtful consideration to their own learning and experience. They are able to assess and understand their strengths and limitations in order to support their learning and personal development.

Approach your exams the IB way

Biology SL

IB DIPLOMA PROGRAMME

Guy Décarie
Jonathan Knopp
David Mindorff

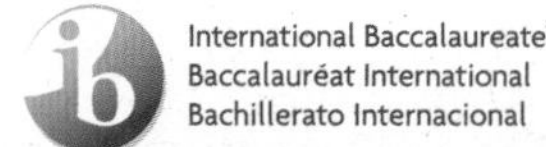

IB Prepared
Approach your exams the IB way
Biology SL

Published January 2011
International Baccalaureate
Peterson House, Malthouse Avenue, Cardiff Gate
Cardiff, Wales GB CF23 8GL
United Kingdom
Phone: +44 29 2054 7777
Fax: +44 29 2054 7778
Website: http://www.ibo.org

The International Baccalaureate (IB) offers three high quality and challenging educational programmes for a worldwide community of schools, aiming to create a better, more peaceful world.

IB merchandise and publications can be purchased through the IB store at http://store.ibo.org. General ordering queries should be directed to the sales and marketing department in Cardiff.

Phone: +44 29 2054 7746
Fax: +44 29 2054 7779
Email: sales@ibo.org

British Library Cataloguing in Publication Data.
A catalogue record for this book is available from the British Library.

ISBN: 978-1-906345-28-0

Cover design by Pentacor**big**
Typeset by Wearset Ltd
Printed and bound in Spain by Edelvives

Item code 4054

2015 2014 2013 2012 2011

10 9 8 7 6 5 4 3 2 1

Acknowledgments
Alison Davies and Andrew Allot for advice on IB Diploma Programme biology standard level

Table of contents

1. Introduction

As an IB student, you are provided with many resources on the path to your final biology standard level (SL) exams. This book is just one of the resources to help you prepare.

How to use this book

Welcome to IB Prepared! This book contains advice and information to be used by students throughout the year. It cannot replace any of the content material that you are expected to have learned before taking your exams.

The main function of this book is to help you prepare for your exams. It includes advice on how to approach exam questions. Actual IB student answers are accompanied by senior examiner comments highlighting how marks were gained and suggestions for how students might have improved their answers. We suggest that, before you read the students' responses, you try to answer the questions by yourself.

Whenever you review student answers found in the book, you will find that green ticks (✓) indicate where marks have been earned. Red numbers (for example, ❶) and green numbers (for example, ❷) are often linked to comment boxes that discuss the areas for improvement and strengths in the student response. Blue boxes contain a summary statement about the positive qualities in the answer. Because positive marking is used, you will find example answers where students have earned maximum marks even though their answers can be improved upon. This is why top answers may show red boxes.

What is in this book?

- **Chapter 2** focuses on the structure of the exam papers.
- **Chapter 3** explains the command terms. These are key words used in exam questions that tell you what is expected in your written answers.
- **Chapter 4** provides top tips. These are bits of general advice and strategies that have been created as a result of reviewing the reports written by examiners (the people who mark your exams).
- **Chapter 5** provides suggestions for how to prepare for studying.
- **Chapter 6** provides advice and strategies for approaching quantitative exam questions, including data-based questions.
- **Chapters 7–12** have been broken up to match the syllabus topics you are taught.
  - The first half of each chapter contains sections of syllabus topics. Each one starts with definitions of important terminology from that section of the syllabus and then identifies key concepts (through "You should know" and "You should be able to") as well as common pitfalls in a section entitled "Be prepared". These sections of syllabus topics are backed with worked examples from a mix of questions chosen from papers 1 and 2 to show you the range of possible questions.
  - The second half of each chapter gives further examples of the questions you may face in your exams, with actual student answers. For each answered question, we provide examples of a range of performance. These are actual answers, written by IB students like you. The exam questions are accompanied by suggestions to help you approach the question and should only be used as guidance. To help you gain an insight into the way examining works, we provide you with the mark that was given for each answer and an explanation, written by a senior examiner, of strengths and weaknesses in the student response.
- In **chapters 13–19**, we follow a similar format, but with questions specific to the topic options, taken from paper 3.
- Last, in **chapter 20**, you will find further examples of exam questions, allowing you to put into practice what you have learned from this book.

2. Get to know your exam papers

This chapter will explain the structure of the exam papers to give you an overall view.

Overview of the exam papers

Your exam consists of three separate papers given to you over two days.

You will write paper 1 and paper 2 in the afternoon of the first day and paper 3 in the morning of the second day.

The exam papers are designed to assess the degree to which you are able to:

- display knowledge of factual information from the syllabus
- apply and use concepts and principles
- construct explanations of biological phenomena
- make predictions
- select and apply relevant information, concepts and principles in a variety of contexts
- analyse and evaluate quantitative and/or qualitative data
- solve quantitative and/or qualitative problems
- communicate logically and concisely
- use appropriate terminology and conventions
- show insight or originality.

Over the three exam papers, there is a range of question types:

- multiple-choice (found in paper 1)
- structured data analysis (found in papers 2 and 3)
- structured factual recall (found in papers 2 and 3)
- extended response (found in papers 2 and 3).

Component	% of marks*	Objectives (% marks)		Time	Details	
		1 + 2	3			
Paper 1, SL	20	20		45 minutes	• 30 multiple-choice questions	
Paper 2, SL	32	≈16	≈16	1 hour 15 minutes	• Section A: ▪ one data-based question ▪ several short-answer questions on the core (all compulsory)	• Section B: ▪ one extended-response question on the core (from a choice of three), each question subdivided into three parts (a, b and c) with a more or less common theme
Paper 3, SL	24	≈12	≈12	1 hour	In each of the two options studied: • one data-based question • several short-answer and/or explanation questions (all compulsory)	

* **Note:** Internal assessment accounts for the remaining 24% of the final mark.

Paper 1 details

Outline

- 30 multiple-choice questions worth a total of 30 marks.
- Each is stated as a question.
- Covers topics 1–6 (core).
- 45 minutes in length, with no reading time.
- Use of a calculator is **not** permitted.
- Use of a simple translation dictionary is permitted if English is not your first language.
- Most of the questions will be simple completion questions involving a stem with three distractors and one correct answer. (See example 1 below. Correct answers are circled in the examples below.)
- Some of the questions will be multiple-completion questions: a stem and three options, of which one, two or three may be correct. (See examples 2 and 3 below.)
- Questions may contain text only or text and figures/diagrams. (See example 4 below.)
- Questions may stand alone or a single stem might apply to more than one question (see example 5 below.)
- Possible answers may be images (see example 6 on page 4) or may be presented in a table (see example 7 on page 4).

Example 1: Simple completion question

Which of the following is a characteristic of Platyhelminthes?

A. Many pairs of legs
B. Flat body
C. Hard exoskeleton
D. Presence of cnidocytes

Most questions are straightforward, offering only a choice of answers.

Example 2: Multiple-completion question

Which of the following structures are present in **both** plant and animal cells?

I. Cell wall
II. Chloroplast
III. Mitochondrion

A. I only
B. I and II only
C. I and III only
D. III only

Example 3: Multiple-completion question

Which of the following occur(s) at birth in the mother's body?

I. Increase in oxytocin
II. Increase in uterine contractions
III. Increase in levels of progesterone

A. I only
B. I and II only
C. II and III only
D. I, II and III

In examples 2 and 3, three different statements are given and the choice of answers is a combination of these statements. Statements are identified with roman numerals, whereas answers are always identified with letters. There is always only one valid answer identified with letters.

Example 4: Multiple-choice question accompanied by an illustration

Which diagram represents the polarity of a water molecule?

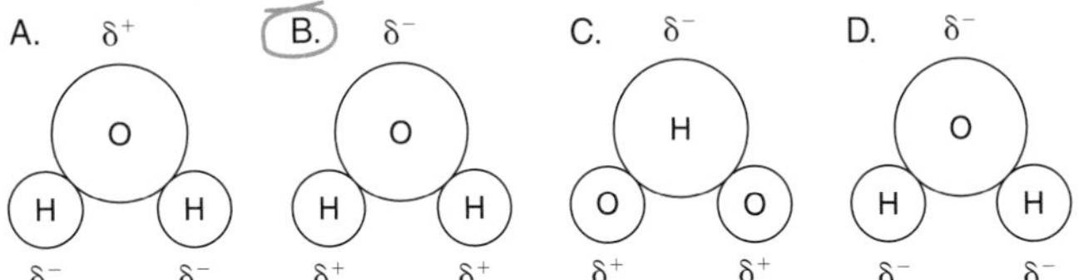

Example 5: Multiple-choice question with more than one question tied together by a stem

The following diagram of a prokaryote refers to questions 1 and 2.

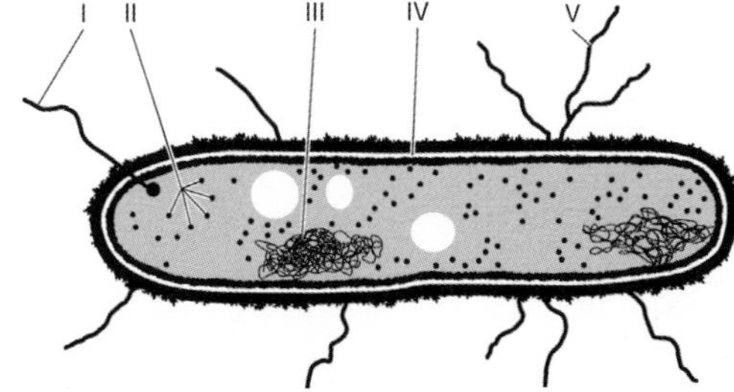

1. What is the function of structure II?
   A. Passing of hereditary information to offspring
   B. Movement of the organism
   C. Regulation of the entry and exit of materials
   D. Production of proteins

2. Which structures are found in all eukaryotic and prokaryotic cells?
   A. I and II only
   B. II and IV only
   C. II and V only
   D. III and V only

Example 6: Multiple-choice question where possible answers are images rather than text choices

Which graph shows the sigmoid population growth curve?

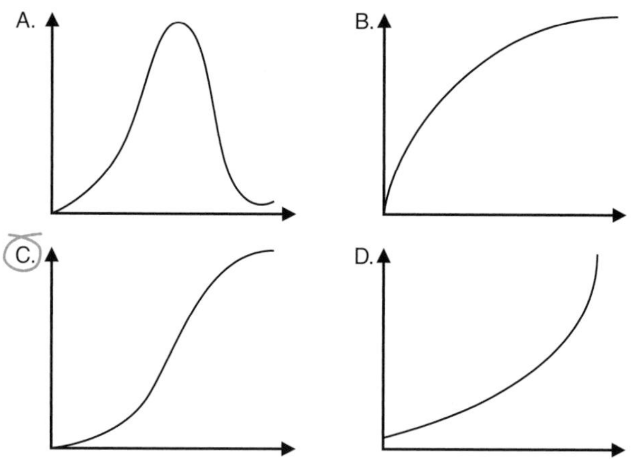

Example 7: Multiple-choice question where possible answers are presented in a table

What is the difference between simple diffusion and facilitated diffusion?

	Simple diffusion	Facilitated diffusion
A.	Rate decreases with increasing concentration gradient	Rate increases with increasing concentration gradient
B.	Faster movement of molecules	Slower movement of molecules
C.	Always involves a membrane	Never involves a membrane
D.	Uses any part of a membrane	Uses channels in the membrane

Paper 2 details

Outline

- This paper involves written responses that expect you to be familiar with command terms.
- 50 marks.
- 1 hour and 15 minutes, not including 5 minutes reading time.
- Section A: 30 marks, with no choice.
- Question 1 involves questions structured around data analysis (data-based questions, DBQ).
- The remainder of the questions in section A involve questions structured around factual recall.
- Section B is out of 20 marks and involves extended response questions. You must choose one question out of three, each one being graded out of 20 marks.
- Use of a graphic display calculator is permitted.

Section A data-based questions

For papers 2 and 3, there are **data-based questions**. These questions involve a piece of data, such as a chart, graph, diagram, electron micrograph, table, and so on, that you will probably have never seen before, often taken from the scientific literature. Usually, in paper 2, there are up to three pieces of data within a question, with a common theme, whereas, in paper 3, there is often only one piece in each option. Each of these questions is divided into sub-questions (a, b, c, etc). There are usually some sub-questions directly on the data, where you would be asked to do such tasks as measure a value, outline a trend, calculate a percentage change or compare a set of values. Other sub-questions may follow that relate to syllabus content including the evaluation of hypotheses (see part (c) in example 8 below) or the application of the knowledge (see part (d) in example 8 below).

Example 8: An example of a data-based question

The symptoms of asthma vary according to the time of year. A study was carried out in New York to determine if an increase in the amount of pollen in the air caused an increase in the number of asthma attacks. Over a period of 270 days, the number of people admitted to New York hospitals with asthma attacks was recorded. The graph below shows this data together with the pollen count.

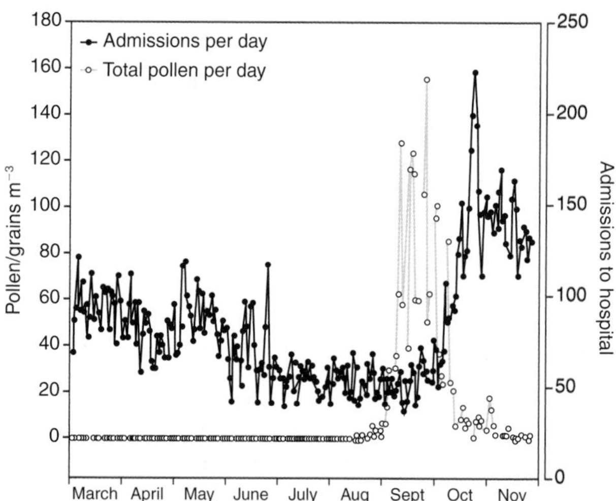

[*Source*: Jamason, PF et al. 1997. "A synoptic evaluation of asthma hospital admissions in New York City". *American Journal of Respiratory and Critical Care Medicine*. Vol 156. Pp 1781–1788. Official journal of the American Thoracic Society. Reprinted with permission from the American Thoracic Society. Copyright © American Thoracic Society. Diane Gern, Publisher.]

(a) Identify the greatest number of hospital admissions in one day. *[1]*

(b) Describe how the numbers of hospital admissions changed over the period of study. *[3]*

(c) Evaluate whether the hypothesis that pollen in the air increases asthma attacks is supported by the data. *[2]*

(d) Asthma causes narrowing of the bronchioles. List two structures that link to bronchioles. *[2]*

Section A simple factual recall question

These questions are used to evaluate knowledge and understanding of the syllabus content. They are often divided into parts (a), (b) and so on, and parts may or may not share a common theme. Most of the questions will be for 1, 2 or 3 marks. Factual recall is what is being assessed, and you can usually answer in the few lines provided.

Example 9: Section A simple factual recall question

(a) Outline the bonding between DNA nucleotides. *[2]*

(b) Explain how chemical bonding between water molecules makes water a valuable coolant in living organisms. *[2]*

(c) State a word equation for anaerobic cell respiration in humans. *[1]*

Section A structured factual recall question

These questions are used to evaluate knowledge and understanding of the syllabus content. They are often divided into parts (a), (b) and so on. Most of the questions will be for 1, 2 or 3 marks. Sometimes there might be a short bit of reading or some data might be used. However, factual recall is what is being assessed, not data analysis.

Example 10: Section A structured factual recall question

A gene in humans called APC is located on chromosome 5. This gene controls cell division and is known as a tumour suppressor gene. Mutations of APC cause a genetic disease called FAP (familial adenomatous polyposis).

(a) State, with a reason, whether FAP is a sex-linked genetic disease or not. *[1]*

50% of the gametes produced by a person with FAP have an APC gene with the mutation.

(b) Identify, with a reason, whether FAP follows a dominant or recessive pattern of inheritance. *[2]*

In a person with FAP, each cell contains a copy of the APC gene with the mutation. If a mutation occurs on the cell's other copy of the APC gene, the cell becomes a tumour cell. Almost everyone with FAP develops cancer before the age of 50.

(c) Explain why almost everyone with FAP eventually develops cancer. *[2]*

In 2004, doctors in Britain were given permission to test embryos to see whether an APC gene with the mutation is present. This test can be used where one of the parents is known to have FAP. The procedure involves the parents using in-vitro fertilization (IVF) to produce embryos, testing the embryos for the gene and implanting only embryos that do not have the mutation.

(d) (i) State the name of this type of test. *[1]*

(ii) State **one** advantage and **one** disadvantage of testing embryos in this way. *[2]*

Advantage:

Disadvantage:

Section B

- 20 marks total.
- Each question will be divided into three parts drawn together by a common theme. Two marks are for the quality of construction of the answer (explained in chapter 4), leaving 18 for content.
- One of the three parts will be longer (about 8 marks).
- The other two parts could be for 4 and 6 marks or 5 and 5 or some other combination to make up the balance of 18 marks.
- They will be as broad as possible in terms of coverage of the syllabus.

Example 11: Part B extended-response question

(a) Outline the role of hydrolysis in the relationships between monosaccharides, disaccharides and polysaccharides. *[4]*

(b) Describe the use of biotechnology in the production of lactose-free milk. *[6]*

(c) Explain the importance of enzymes to human digestion. *[8]*

Paper 3 details

Outline

This paper involves written responses that expect you to be familiar with command terms.

- 36 marks.
- 1 hour.
- Choose two options from seven according to what has been taught at your school.
- 18 marks each.
- Each option contains several short-answer questions and one extended-response question.
- Programmable/graphic display calculators are permitted for this paper.
- Each option usually consists of three questions, which can be subdivided into (a), (b) and so on, and further subdivided into (i), (ii) and so on.
- One of the questions should include some data (one piece of data is the norm, but there might be two).
- The data could be in the form of a table, bar chart, column graph, histogram, pie chart, line graph, scatter graph and so on.
- The data will be relevant to the option, although not necessarily covering a concept related to any of the option assessment statements.
- The second question will usually carry 6–8 marks and will consist of themed/linked smaller part questions that aim to cover as much of the syllabus as possible.
- The third question will usually be out of 6 marks to test more syllabus coverage. It may be subdivided or presented as a single question involving a more extended response.

Cover sheets, materials and stationery

Cover sheets

Students have personalized cover sheets to attach to each paper (except for multiple-choice question papers, where computer readable sheets are used). We have reproduced a full cover sheet here.

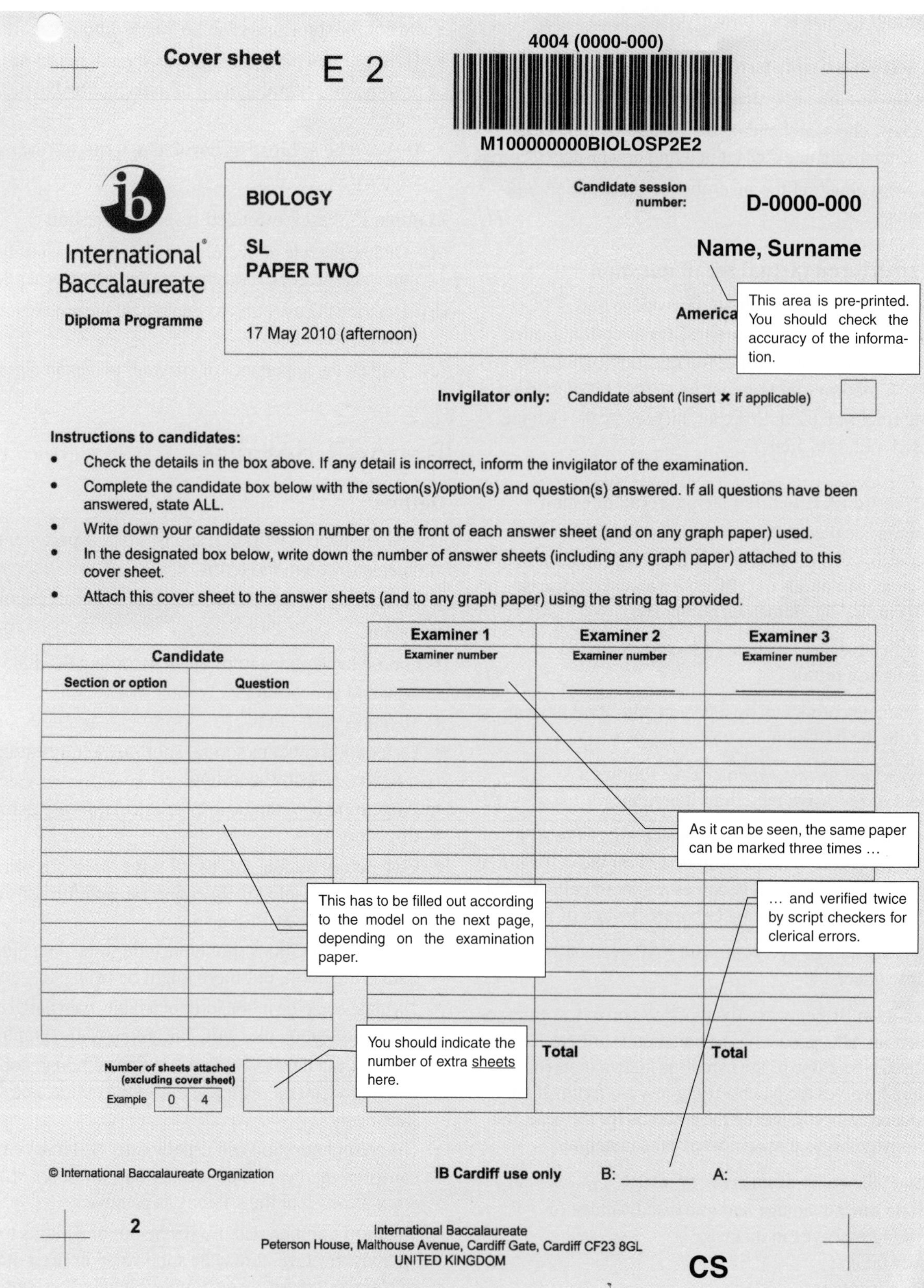

Cover sheet E 2

4004 (0000-000)

M100000000BIOLOSP2E2

International Baccalaureate

Diploma Programme

BIOLOGY

SL

PAPER TWO

17 May 2010 (afternoon)

Candidate session number: D-0000-000

Name, Surname

America

Invigilator only: Candidate absent (insert ✖ if applicable)

Instructions to candidates:

- Check the details in the box above. If any detail is incorrect, inform the invigilator of the examination.
- Complete the candidate box below with the section(s)/option(s) and question(s) answered. If all questions have been answered, state ALL.
- Write down your candidate session number on the front of each answer sheet (and on any graph paper) used.
- In the designated box below, write down the number of answer sheets (including any graph paper) attached to this cover sheet.
- Attach this cover sheet to the answer sheets (and to any graph paper) using the string tag provided.

Candidate	
Section or option	Question

Number of sheets attached (excluding cover sheet)

Example 0 4

Examiner 1	Examiner 2	Examiner 3
Examiner number	Examiner number	Examiner number
	Total	Total

© International Baccalaureate Organization

IB Cardiff use only B: ______ A: ______

2

International Baccalaureate
Peterson House, Malthouse Avenue, Cardiff Gate, Cardiff CF23 8GL
UNITED KINGDOM

CS

The models below show what you are expected to write under "candidate" for a biology paper 2 and paper 3 (expectations for different subjects are different). In the box "Number of sheets attached", what is meant is the actual number of extra loose sheets of paper (not pages) that you have used to complete the answers or to answer essay questions—the exam booklet does not count, nor the cover sheet. The idea is to figure out if something is missing.

Paper 2 (SL)

Candidate	
Section or option	**Question**
A	all
B	4

Paper 3 (SL)

Candidate	
Section or option	**Question**
E	all
G	all

Materials and stationery

For all papers, your school should provide rough paper for your notes. Be aware that rough paper is not sent to the examiner.

For **paper 1**:

- your school should provide a pencil
- the IBO provides a personalized answer sheet
- verify that all pre-printed details are correct and read the printed instructions.

For **papers 2 and 3**, you should receive:

- an IB personalized cover sheet
- an exam booklet containing the questions
- lined answer sheets
- a treasury tag (string to attach papers).

For **papers 2 and 3**, you are allowed to have:

- a graphic display calculator
- geometry instruments (ruler, square, protractor).

If you need clarifications about exam procedures, consult your IB Diploma Programme coordinator.

Future electronically marked papers

All Paper 2 and Paper 3 will be marked electronically (e-marked) in the near future. In these papers, each short-answer question will be followed by a box containing dotted lines (or a space for drawing). Also, for Paper 2 part B, sheets with dotted lines contained in a box will be printed in the examination booklet.

You should answer on the lines with black ink. If absolutely necessary, you can continue writing below the lines, but **be aware that all material written outside the boxes will not be marked**, as the examiner will not be able to see it.

3. Command the command terms

This chapter explains the command terms. These are key words used in exam questions that tell you what is expected in your written answers.

Paper 2 and paper 3 exam questions are classified as objective 1, 2 or 3 according to the command terms used. The command terms indicate the depth of treatment required for a given assessment statement. For this reason, it is important that you are aware of the meanings of the command terms.

Objective 1 command terms

Define	Give the precise meaning of a word, phrase or physical quantity.
Draw	Represent by means of pencil lines.
Label	Add labels to a diagram.
List	Give a sequence of names or other brief answers, with no explanation.
Measure	Find a value for a quantity.
State	Give a specific name, value or other brief answer without explanation or calculation.

Objective 2 command terms

Annotate	Add brief notes to a diagram or graph.
Apply	Use an idea, equation, principle, theory or law in a new situation.
Calculate	Find a numerical answer, showing the relevant stages in the working (unless instructed not to do so).
Describe	Give a detailed account.
Distinguish	Give the differences between two or more different items.
Estimate	Find an approximate value for an unknown quantity.
Identify	Find an answer from a given number of possibilities.
Outline	Give a brief account or summary.

Objective 3 command terms

Analyse	Interpret data to reach conclusions.
Comment	Give a judgment based on a given statement or result of a calculation.
Compare	Give an account of similarities and differences between two (or more) items, referring to both (all) of them throughout.
Construct	Represent or develop in graphical form.
Deduce	Reach a conclusion from the information given.
Derive	Manipulate a mathematical relationship(s) to give a new equation or relationship.
Design	Produce a plan, simulation or model.
Determine	Find the only possible answer.
Discuss	Give an account including, where possible, a range of arguments for and against the relative importance of various factors, or comparisons of alternative hypotheses.
Evaluate	Assess the implications and limitations.
Explain	Give a detailed account of causes, reasons or mechanisms.
Predict	Give an expected result.
Show	Give the steps in a calculation or derivation.
Solve	Obtain an answer using algebraic and/or numerical methods.
Suggest	Propose a hypothesis or other possible answer.

4. Top tips

Here are some top tips—general pieces of advice and strategies that have been created as a result of reviewing the reports written by examiners (the people who mark your exams).

Paper 1 strategies

- Paper 1 is the only paper where five minutes reading time is not given at the start of the exam.
- However, there is sufficient time to scan the paper to identify challenging questions.
- The introductory text of a question is called a "stem". Read the stem of each question carefully.
- Underline key words.
- Rephrase the question into your own words.
- Watch for small but important words, such as "not" and "always".
- One approach is to cover up the alternatives before you read the stem (the first part of the question) and predict an answer.
- Uncover the alternatives and read all of them carefully, even if the first choice seems correct.
- Eliminate choices you know not to be correct.
- Be careful, but do not read too much into the question.
- Identify the best response.
- If you are not sure of an answer, mark the question on the exam booklet and keep going. Return to this question once you have completed the easier questions.
- There is no additional penalty for incorrect answers, so do not leave any unanswered questions.
- You will be given an answer sheet with your name on it. Record each of your answers with an X according to the instructions printed on the answer sheet.
- Check the answer sheet to be sure you have written an answer for each question. The paper is positively marked, so you will not lose marks for wrong answers. Review the answer sheet to make sure that you have put your answers in the intended spaces.

Paper 2 and paper 3 strategies

- A ruler is required, as is a graphic display calculator.
- When only a specific number of responses are requested, do not include extra responses.
- If no specific number is requested, then do write as much relevant material as you can. A mark scheme often has more specific points worthy of a mark than the total allows. This is intentional.
- A mark will often be deducted if the unit or the percentage symbol is left off. If there is a graph, the unit is often part of the label on one axis of the graph.
- Measurements made with a ruler need to be precise.
- Chapter 6 has specific strategies for answering data-based questions.

- Consider making a rough Venn diagram (a diagram that symbolically shows conceptual areas of overlap using circles) or a table before answering **compare** questions. Always ensure that you have included similarities and differences for each point in your comparison.
- Consider making a table to answer a **distinguish** question.
- In questions where you must **explain**, it is often important to be using the word "because".
- **Suggest** questions require you to use the evidence provided to justify one possible answer.
- **Discuss** questions require you to give an account including, where possible, a range of arguments for and against the relative importance of various factors, or comparisons of alternative hypotheses.
- **Evaluate** questions require you to assess the implications and limitations. Many paper 2 data-based questions require you to evaluate whether the data supports a particular hypothesis.
- **Analyse** questions require you to interpret data to reach conclusions. State as many conclusions as possible.

Communication strategies

The accurate use of correct terms and the correct interpretation of command terms are important tools for success.

- When a question includes a biological term, define the term in your answer.
- Do not leave ideas unstated. Sometimes ideas that you believe are too obvious should be included in your answer to be sure that you communicate all that you know.
- Do not write sentences with too many ideas compacted in them. Separate distinct ideas into different sentences.
- Your outline or explanation of processes or mechanisms must be stated clearly and cannot be substituted by the sole use of examples. Use examples to back-up your answer instead.
- Include the names of relevant organisms in your examples. This is what is meant when you are asked to give a "named" example. Try to be as specific as possible, even genus and species if you can.
- Take care with your spelling as many biological terms with very different meanings are spelled in a similar way.
- Do not confuse words that may sound alike. For example, many students confuse afferent with efferent, affect with effect, and glycogen with glucagon.
- Do not start off structured recall questions by rewriting the question. Look at the number of marks allocated to determine the number of distinct ideas required.

For extended-response questions, up to 2 marks are awarded for the quality of construction of the answers. Here are the directions given to examiners about the quality of construction marks.

"Two aspects are considered:

- expression of relevant ideas with clarity
- structure of the answers.

  One quality mark is to be awarded when the candidate satisfies **each** of the following criteria. Thus **two** quality marks are awarded when a candidate satisfies **both** criteria.

  Clarity of expression:
  The candidate makes a serious and full attempt to answer all parts of the question and the answers are expressed clearly enough to be understood with little or no re-reading.

  Structure of answer:
  The candidate has linked relevant ideas to form a logical sequence **within** at least two parts of the **same question** (for example, within part a and within part b, or within part a and within part c, etc, but **not between** part a and part b, or between part a and part c, etc)."

Drawing strategies

- There should be no gaps in the lines when drawing closed shapes such as cells or mitochondria.
- Unambiguous labels should be included. This is related to students making drawings that are unnecessarily small. The size of the drawing should be proportional to its complexity. It is difficult to distinguish detail or to identify what a label is pointing to in a very small drawing.
- There should be correct positioning of structures in relation to other structures, such as in the correct position of attachment of molecules within a nucleotide.
- Connections between structures should be clearly and correctly shown, such as where the gall bladder connects to the bile duct.
- There should be correct proportions in relation to other structures, such as in not having the mitochondrial cristae drawn too thickly or incorrectly proportioned bars of the different trophic levels in a pyramid of energy.
- As some papers are to be electronically marked, you should not draw on graph paper or use coloured pens, pencils or highlighters, which can obscure the answers. Write in black ink and use black pencil to draw. It is more difficult to read poorly written or poorly drawn answers, so try to be neat.
- In future e-marked papers, boxes will be provided in the examination booklet for your drawings. You should draw within the box. **Be aware that all material drawn or written outside the box will not be marked**, as the examiner will not be able to see it.

Example question

Draw a labelled diagram of the heart showing the chambers, associated blood vessels and valves. *[4]*

How do I approach the question?

- Right and left ventricles should be distinctly separate from each other and from the atria. Atria should be shown as smaller than the ventricles.
- The walls of the atria should be thinner than the walls of ventricles.
- The right ventricle should have thinner walls than the left ventricle.
- AV valves as well as semilunar valves should be correctly located. You can use the terms tricuspid and bicuspid valves. The aorta and pulmonary artery should be shown leaving the appropriate ventricle with semilunar valves shown. The pulmonary vein and vena cava should be shown entering the appropriate atrium.

Which areas of the syllabus is this question taken from?

- The transport system (6.2.1)

This answer achieved 1/4

1. The pulmonary vein and the venae cavae are shown entering the correct chambers.

1. The ventricles are shown as the same size as the atria.
2. The left wall of the ventricle should be thicker on the left side.
3. The bicuspid and tricuspid valves are on the reverse side.
4. The pulmonary vein and artery are both entering the left atrium.
5. The pulmonary artery is mislabelled (as coronary artery) so the semilunar valve mark cannot be awarded.
6. The atria and ventricles are not labelled.

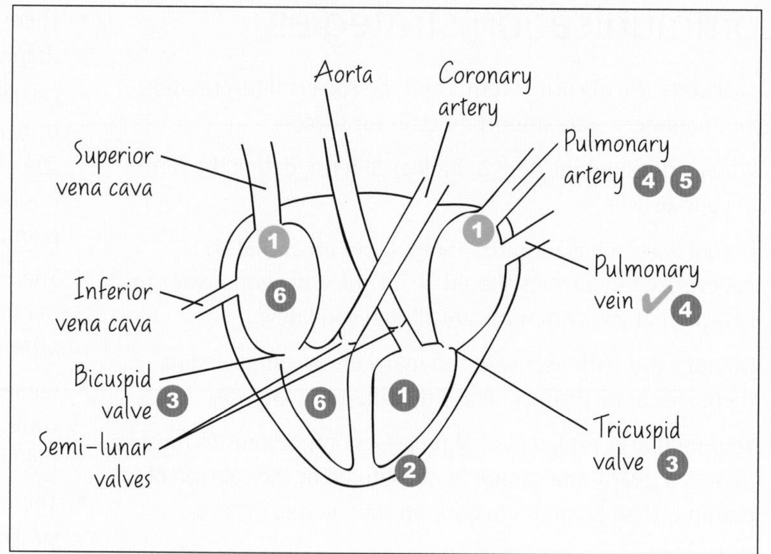

This answer achieved 2/4

1. The left ventricle is shown as thicker than the right ventricle.
2. The atria are shown with thinner walls than the ventricles.

1. The diagram is on the small side for so many accompanying structures and labels.
2. The right atrium and ventricle are shown as the same size.
3. "Atriovascular" is not the correct term.
4. The pulmonary artery is shown exiting the right atrium.
5. The pulmonary vein is shown entering the right atrium.
6. The semilunar valves are not labelled on both sides.

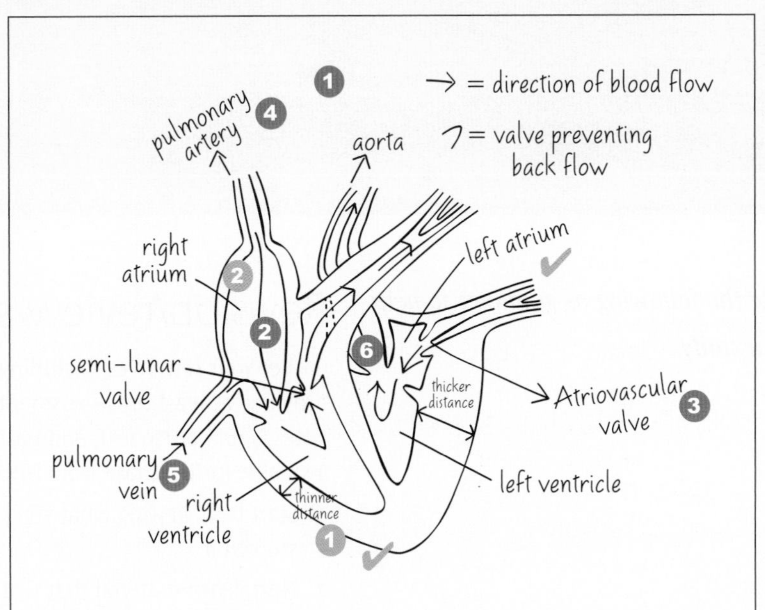

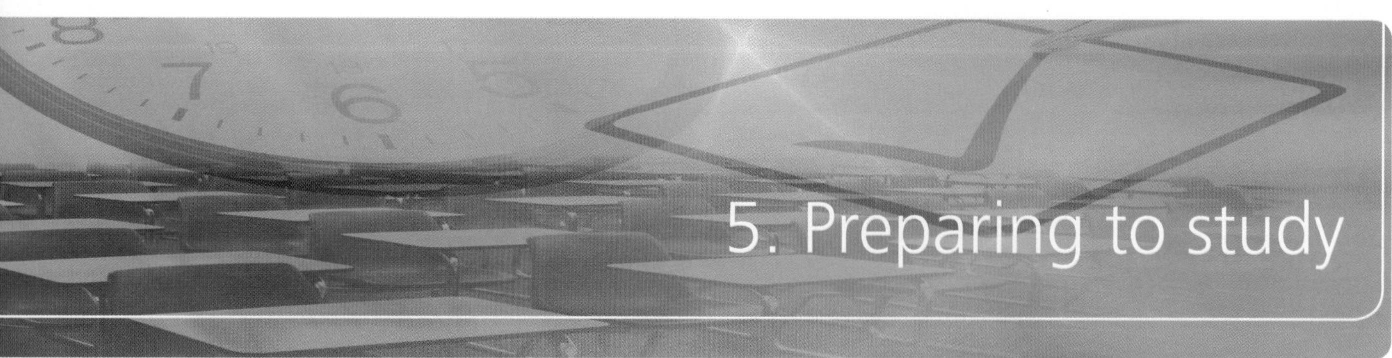

5. Preparing to study

Use the following as guidance to help you study.

Revision/review strategies

Gather your resources, including the subject guide (your teacher/school should have provided at least the assessment statements contained in the subject guide), your notes, your course text, and past papers and their markschemes. Past papers are available for purchase from http://store.ibo.org.

- Start by reviewing what you knew well earlier but may have forgotten since you last studied it.
- Next, learn material that is straightforward but that you never found the time to learn solidly up until now.
- Do some past multiple-choice papers and consult the answers as a diagnostic tool to determine your areas of strength and weakness. Remember that only multiple-choice questions from May 2009 onwards are relevant for the current syllabus.
- Begin to revise the syllabus comprehensively, point by point.
- Study the syllabus as broadly as possible because you cannot anticipate which areas will be covered by the paper.
- Make sure that you have thoroughly studied the options that were covered in class. It is highly dangerous to attempt other options for any reason such as interesting data.
- Meet with your teacher to get help with topics that have confounded you in the past: classical genetics, the nerve impulse, muscles, photosynthesis and respiration are some common examples of such topics.
- Practise paper 2, part B, long answers, and compare your answers to the markschemes.
- Practise part A, short answers.
- Try answering at least two past papers in full.

Interpreting a markscheme

If you have old exams available, you also probably have access to markschemes. The intended users of the markscheme when it was written were examiners, not students.

Consider the following part of a markscheme (the task was to "Describe how skin and mucous membranes act as barriers to infection"):

skin is a physical barrier/impermeable;
skin has an acidic pH which inhibits growth of (pathogenic) bacteria;
outer layer flaked off (to remove parasites/bacteria);
mucous membranes trap microorganisms;
lysozyme secreted and destroys bacteria;
cilia push mucus containing bacteria out;
other/beneficial bacteria on skin keep other pathogens in check; ***[5 max]***

- Each marking point begins on a separate line and the end is signified by means of a semicolon (;). An alternative answer or wording is indicated in the markscheme by a "/". Either wording can be accepted.
- Words in brackets (...) in the markscheme are not necessary to gain the mark.
- The order of points does not have to be as written (unless stated otherwise).
- If your answer has the same "meaning" or can be clearly interpreted as being the same as that in the markscheme, then the mark will be awarded. If a word is underlined in the markscheme, then the word must appear in your answer to be awarded the mark.
- Marking is positive, that is students are given credit for what they have achieved, and for what they have correct, rather than being penalized for what they have not achieved or what they have wrong.

  Consider the following line from an older markscheme:

  41 mg C fixed $m^{-3} d^{-1}$ (±2);
  (*units required for the mark*)

- Units should always be given where appropriate.
- Omission of units will only be penalized once per paper.
- You will not be penalized for errors in significant figures, unless it is specifically referred to in the markscheme.
- Sometimes a range of acceptable answers is allowed, and in an older markscheme will have a plus or minus sign, such as 41 ± 2 in the answer above. This is a note to the examiner to accept answers between 39 and 43. When writing down your numerical values to exam questions, it is not necessary to include an uncertainty figure.

6. Statistical analysis, mathematical requirements and data-based questions

Statistical analysis

You should know:

- the use of the mean, standard deviation and error bars
- the standard deviation summarizes the spread of values around the mean
- 68% of values fall within one standard deviation of the mean
- about causality and correlation.

You should be able to:

- calculate the mean and standard deviation of a set of values with a calculator
- explain how the standard deviation is useful for comparing the means and spread of data between two or more samples
- deduce the significance of the difference of two sets of data using *t*-values
- explain that the existence of a correlation does not establish causality
- transpose statistical skills between syllabus content and internal assessment.

Examples

1. What does the size of the standard deviation indicate about data?

   A. How accurately the data were measured

   B. How widely the data are spread above and below the mean

   C. Whether the mean is larger or smaller than it should be

   D. Whether the reliability of the data is greater or less than 68%

   The standard deviation (SD) is a measurement of the spread of data around the mean, therefore B is the right answer. Although the SD could be used to compare sets of data, it does not indicate the size of the mean (C) nor the accuracy or reliability of data (A and D). The mention of 68% in D is irrelevant.

2. The *t*-test is used to test the statistical significance of a difference. What is that difference?

   A. Between observed and expected results

   B. Between the means of two samples

   C. Between the standard deviation of two samples

   D. Between the size of two samples

   B is the correct answer. The *t*-test is based on the differences between the means of two samples, although the overlap, which could be measured by the standard deviation, is involved (C). The difference between observed and expected results (A) is measured by a χ-square (chi-square, χ^2) test. This is not in the syllabus and is irrelevant here, as well as the size of two samples (D).

Be prepared

- Basic knowledge about the mean and standard deviation may be assessed through paper 1, but the data-based questions in papers 2 and 3 may address the application of this knowledge.

Mathematical requirements

You should be able to:

- perform basic arithmetic functions
- recognize basic geometric shapes
- carry out simple calculations within a biological context involving decimals, fractions, percentages, ratios, approximations, reciprocals and scaling
- use standard notation (for example, 3.6×10^6)
- use direct and inverse proportions
- represent and interpret frequency data in the form of bar charts, column graphs and histograms, and interpret pie charts and nomograms
- determine the mode and median of a set of data
- plot and interpret graphs (with suitable scales and axes) involving two variables that show linear or non-linear relationships
- plot and interpret scattergraphs to identify a correlation between two variables, and appreciate that the existence of a correlation does not establish a causal relationship
- demonstrate sufficient knowledge of probability to understand how Mendelian ratios arise and to calculate such ratios using a Punnett grid
- make approximations of numerical expressions
- recognize and use the relationships between length, surface and volume.

Examples

1. This question refers to the following micrograph of an *E. coli* bacterium undergoing reproduction:

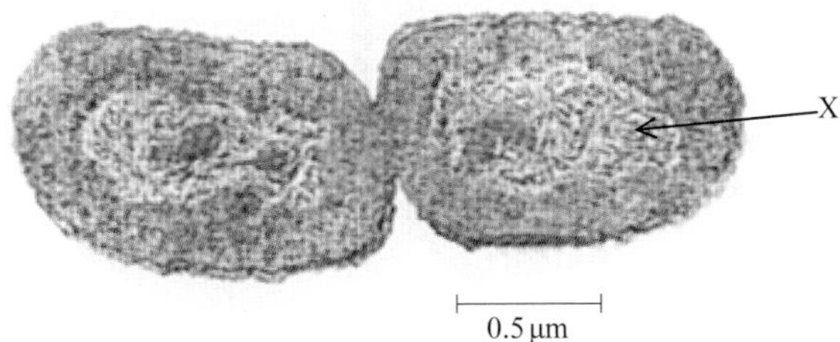

[*Source*: www.bio.mtu.edu/campbell/prokaryo.htm. Copyright © Bedford, Freeman & Worth Publishing Group, LLC.]

The scale bar represents 0.5 µm. How long are both cells in total?

A. 5.0×10^{-6} m

B. 5.0×10^{-9} m

C. 2.5×10^{-6} m

D. 2.5×10^{-9} m

C is the correct answer. Since 1 µm equals 10^{-6} m, choices B and D with 10^{-9} m should not be considered. Divide the length (on paper) of both cells by the length of the scale bar: this establishes a ratio, which is 5 here. Multiply the value of the scale bar by the ratio: $(0.5 \times 10^{-6}\,\text{m}) \times 5 = 2.5 \times 10^{-6}\,\text{m}$.

(Hint: If no ruler is available, copy the ends of the scale bar on the edge of the rough paper provided and use this to estimate the ratio by measuring the length of both cells with it.)

2. An electron micrograph of a liver cell, with scale bar, is shown below. Calculate the magnification of the photograph.

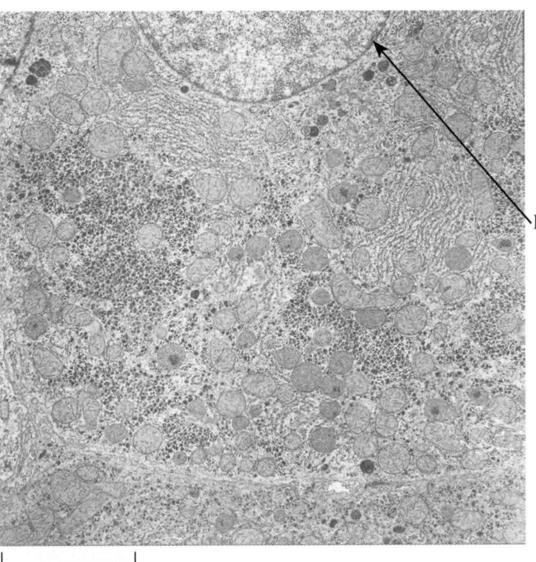

[*Source*: Kalderon, D, Roberts, BL, Richardson, WD and Smith, AE. 1984. "A short amino acid sequence able to specify nuclear location". *Cell*. Vol 39. P. 499. Copyright © Elsevier. Reproduced with permission.]

$$\text{Magnification} = \frac{\text{size of image}}{\text{size of specimen}}$$

The scale bar represents 5 µm. To calculate the magnification, you should measure it with a ruler and convert mm to µm (add three zeros to your measurement in mm).

$$\text{Magnification} = \frac{\text{measurement on paper} \times 10^{-6}\,\text{m}}{\text{value of scale bar} \times 10^{-6}\,\text{m}}.$$

(The measurement on the original exam paper was 30 mm, so the magnification was 6000x, but the image will be reduced in this book, so the actual magnification will be much smaller.)

Mathematical requirements (continued)

Be prepared

- The IB always uses the *Système International* (SI, International System) for units.
- Always show units with your answers—good answers may not receive credit if units are not shown.
- The number of significant digits resulting from a calculation should be consistent with those used in the calculation.
- Skills used to process data and evaluate data in internal assessment are transferable to exam papers.
- Units in exam papers will often be preceded by the symbol "/" and will be expressed using exponential notation. A negative exponent means "divided by" or "per" (for example, "/mg l^{-1}" means "in milligrams per litre").
- Punnett grids to calculate Mendelian ratios in genetics will be used in chapter 9.

Interpreting data

You should be able to:

- interpret data such as a chart, graph, diagram, electron micrograph, table, and so on that are often taken from the scientific literature (papers 2 and 3 have such data-based questions)
- use the skills that you have developed during the internal assessment portion of the course to help with your performance
- interpret unusual data presentation formats such as nomograms, spider graphs and three-dimensional graphs
- measure a value, outline a trend, calculate a percentage change or compare a set of values in some sub-questions
- apply syllabus content to your data interpretation or evaluate a hypothesis (see examples below).

Examples

This is a question on a field study that was conducted among four different species of migrating birds known to stop at high-quality and low-quality food sites where they refuel in order to continue flying. It started with a table of data presenting the number of captured birds and their average mass. After a couple of sub-questions ((a) and (b)), the question continues with the following stem.

A method was used to determine the average mass change in grams per hour ($g\,h^{-1}$) during the study. Graph A represents a summary of data collected during one season whereas graph B represents a summary of data collected over 17 years.

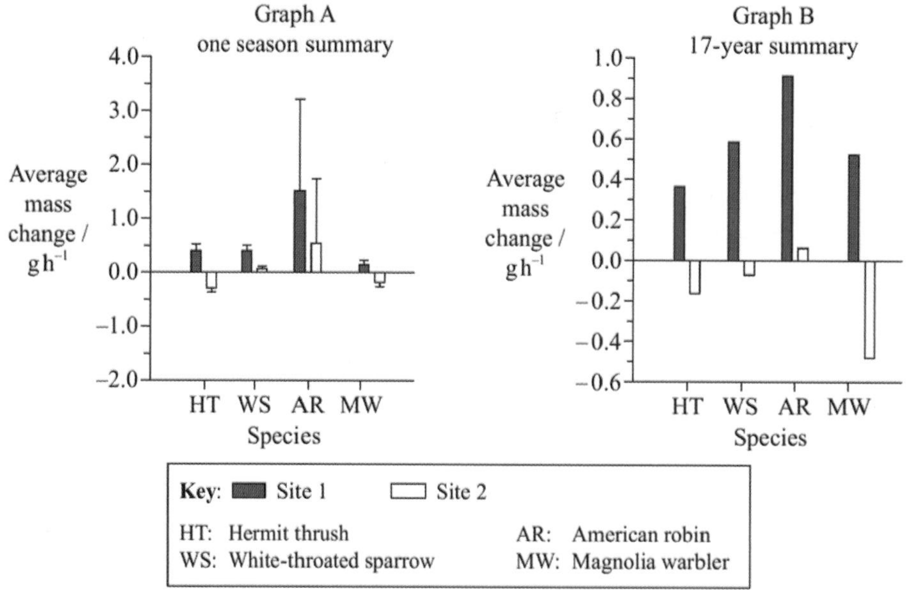

[*Source*: Adapted from Guglielmo, C et al. 2005. "A field validation of plasma metabolite profiling to assess refuelling performance of migratory birds". *Physiological and Biochemical Zoology*. Vol 78, number 1. Pp 116–125.]

Interpreting data (continued)

(c) Compare the 17-year summary data for the hermit thrush and the magnolia warbler. *[2]*

This question requires that you look at graph B only, and at the data for HT and MW only. What examiners want to see is if you have understood the graph: it is pointless to quote values only. You have to realize that all data above 0.0 g h^{-1} represent an increase or a gain in mass, whereas those under 0.0 g h^{-1} represent a decrease or a loss in mass: you must write this in your answer, as "change" was not accepted. Since a comparison is required, there must be at least one similarity: both gain mass at site 1 and lose mass at site 2. There must also be at least one difference: MW has a greater increase at site 1 and a greater decrease at site 2.

(d) Evaluate the one season data for the hermit thrush and the American robin with regard to average mass change per hour at site 1. *[2]*

Here the question asks to "evaluate" the data, not to describe or to explain. In other words, how valid are the data for HT and AR in graph A? Examiners do not want to see values, they want your opinion. The data for HT are reliable, whereas the AR data are not, because of high error bars for AR and its small sample (from the first table of the question, not shown here).

Among birds, high triglyceride concentration in blood plasma indicates fat deposition whereas high butyrate concentration in blood plasma indicates fat utilization and fasting. The following data summarizes triglyceride levels and butyrate levels measured for the same groups of birds.

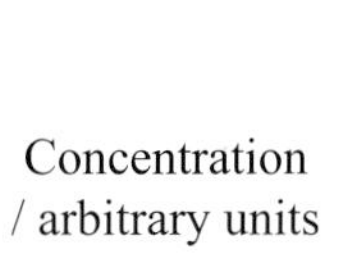

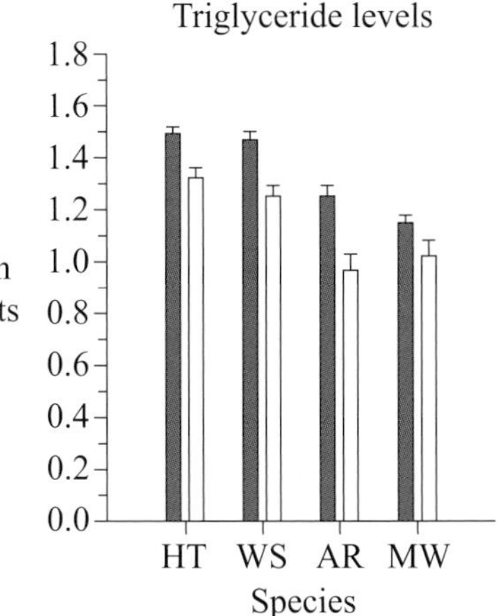

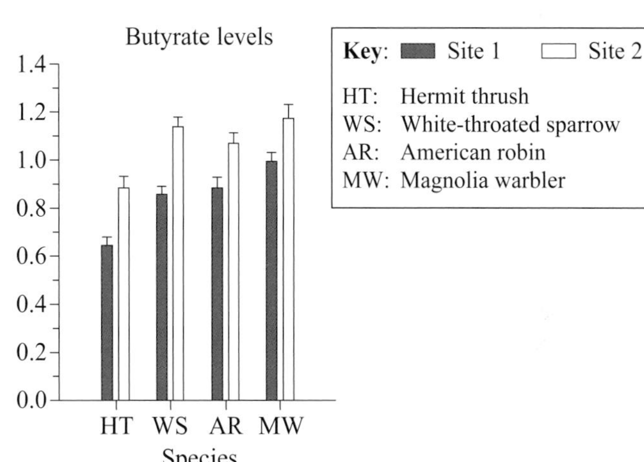

[*Source*: Adapted from Guglielmo, C et al. 2005. "A field validation of plasma metabolite profiling to assess refuelling performance of migratory birds". *Physiological and Biochemical Zoology*. Vol 78, number 1. Pp 116–125.]

(e) Describe, using the triglyceride levels graph, the results at site 1 and site 2 for all of the birds. *[2]*

A description should show the trends, not repeat values. This is also not the place to explain why or how. The question is about the triglyceride graph only, so do not write about the other one. All have a higher concentration of triglycerides at site 1 than at site 2; HT is the highest at both sites; MW is the lowest at site 1; and AR is the lowest at site 2. Your answer should be in terms of triglycerides: this is what is measured by the data. Fat deposition is a consequence of triglyceride concentration and should be mentioned in an explanation (the next question).

(f) Explain the differences in the triglyceride level and butyrate level for the hermit thrush at site 1 and site 2. *[2]*

This is your opportunity to give an explanation. You should have described the data in the previous question, so it is pointless repeating the description here. There is a higher fat deposition at site 1 because HT eats more and has a higher triglyceride level. There is a higher butyrate level at site 2 because HT fasts more and uses more fat or triglycerides.

Interpreting data (continued)

Be prepared

A data-based question often requires you to accomplish different tasks from the same piece of data. These tasks are presented in the different sub-questions following the data and use different command terms.

- Often in questions that ask students to "outline" and "describe", students will inappropriately give answers that deduce a conclusion or explain the results. It is important to know your command terms well and respond appropriately.
- When asked to "outline" a pattern in the data, you should give a big-picture summary of what is going on. Detailed descriptions that quantitatively trace the changes of each group are unnecessary and may, in fact, never include identification of any trends.
- When asked to "describe" a pattern in the data, you should give a big-picture summary but you should also include an element of quantification such as "the mass increased from 190 to 295 units in 85 days but then remains stable".
- "Explain the results" requires that you write the reasons or the mechanisms causing the results. There is then no point quoting values in your answer unless you need them to explain different parts of the data.
- "Suggest" requires that you use your knowledge, sometimes from different areas of the syllabus, to provide or extrapolate one or many possible causes for the results, although this would be only speculative.

Hypothesis analysis

You should be able to:

- recall, from the conclusion and evaluation section of the internal assessment, that data gathered as a result of experiments can have limitations and can support multiple interpretations
- consider the degree to which the data presented in some data-based questions supports a particular hypothesis or conclusion
- assess the implications of a hypothesis as well as any limitations, in any evaluation.

Examples

This question is part of the previous example, but addresses hypothesis analysis.

(g) Scientists have hypothesized that the food quality is better at site 1 than at site 2. Evaluate this hypothesis using the data provided. *[2]*

The question asks you to evaluate the hypothesis, not the data. Is the hypothesis supported, and how? You should use all the data available, unless specified otherwise. The hypothesis is supported because: the mean mass at site 1 is higher than at site 2 for all birds (from the first piece of data, not shown in this example); mass is gained at site 1 and lost at site 2; triglyceride level is higher at site 1, showing fat deposition, whereas butyrate level is higher at site 2, showing fasting and fat utilization. In this case, all data support the hypothesis.

Be prepared

- Many of the skills that you have developed due to the internal assessment portion of the course can help with exam performance.
- In an evaluation, you can look at a range of possibilities including alternative hypotheses if the data support such a discussion.
- You should understand the distinction between a correlation and a cause-and-effect relationship.
- Some common limitations of data are: experiments done in animal models being extrapolated to humans; too few bits of data; and highly variable data.
- "Evaluate the hypothesis" means that you should write how the trends support/do not support the hypothesis. Here, again, there is no point describing the data and/or quoting values. Sometimes you may also have to refer to the previous pieces of data that were presented in the question. Depending on the data provided, sometimes all the evidence clearly supports the hypothesis or it does not. Sometimes, data are contradictory and at other times there are not enough data either to support or not to support the hypothesis. You will have to judge the situation and write which element(s) support or do not support the hypothesis.

There is evidence that body size of animals tends to increase over time. In this study, fossils and living species from the genus *Poseidonamicus*, deep-sea ostracods, were used to test this hypothesis. The numbers on the dotted line represent the number of different *Poseidonamicus* species found either as fossils or living. For each time period, the average valve length of all species studied is plotted. Valve length is an indication of total body size. The continuous line is the estimated temperature of their deep-sea habitat.

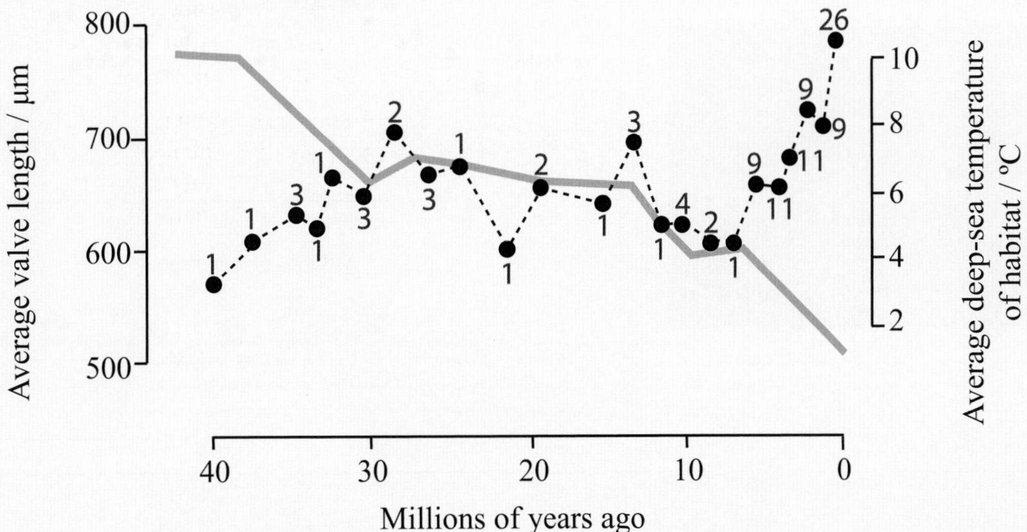

[*Source*: Hunt, G and Roy, K. 2006. "Climate change, body size evolution, and Cope's rule in deep-sea ostracods". *Proceedings of the National Academy of Sciences*, USA. Vol 103. Pp 1347–1352. Copyright 2006 National Academy of Sciences, USA. Reproduced with permission.]

(a) Calculate the percentage increase in valve length between the species studied from 40 million years ago and the species from the present day. *[2]*

[Taken from higher level paper 3, time zone 1, May 2009]

How do I approach the question?

The first step is to find the appropriate values for the calculation, using a ruler. Then the difference must be calculated, divided by the first value, times 100.

The correct calculation is:

$$\frac{790-570}{570} \times 100 = 39\% \text{ (units required)}$$

The subtraction 790 − 570 alone did not give any marks, as the complete calculation (including the division by 570) was required for the first mark. The percentage symbol was required for the second mark.

Also, always remember to look at how units are expressed in exam questions.

Which areas of the syllabus is this question taken from?

- The mathematical requirements of the syllabus

This answer achieved 2/2

1. The calculation is correct.
2. An acceptable value is found (range 37–41%) and the units are written.

1. The student did not measure using a ruler, which could have led to inaccuracies and an answer out of the accepted range.
2. The first used value is incorrect (it should be 790 µm).
3. Too many significant digits were used, but this was not penalized.

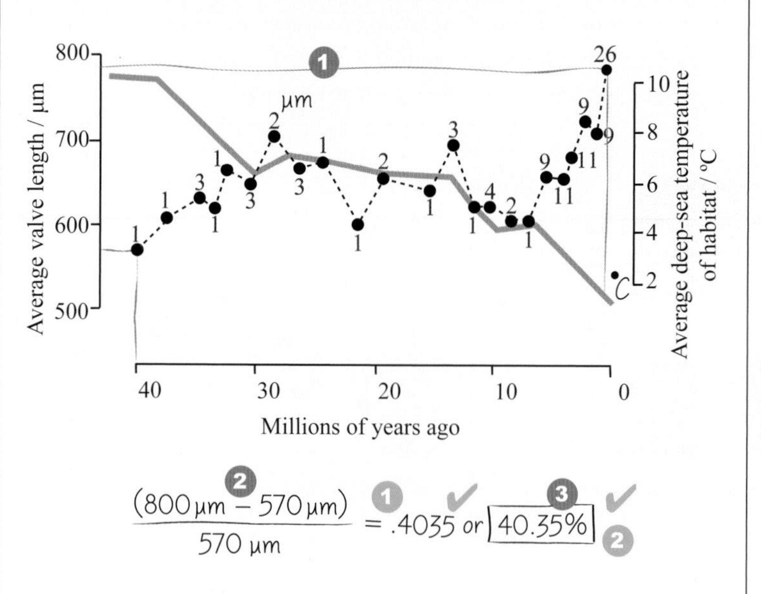

[This question refers to the data in the previous example.]

Evaluate the hypothesis that changes in size of *Poseidonamicus* are caused by changes in sea temperature. *[3]*

How do I approach the question?

An evaluation requires students to assess implications and limitations. In this case, there does appear to be a correlation and therefore support for the hypothesis, but there are a number of components to the data that limit the degree to which we can be certain of conclusions. While the data does support a relationship between valve length and temperature, the valve length is quite variable during a time when sea surface temperature was relatively stable.

Which areas of the syllabus is this question taken from?

- Aim 4 of the syllabus, "develop an ability to analyse, evaluate and synthesize scientific information"

This answer achieved 1/3

1. The student is describing the data rather than evaluating the hypothesis.
2. An explanation is not expected when the command term is "evaluate".

There is a weak inverse correlation between size and sea temperature. ✔ At 40 million years ago, the sea was 10°C and the size was 575 µm. Today the sea is 2°C and the length is 800 µm. (1) This may be because the changing sea temperature favoured some of the organisms over others. (2)

This answer achieved 3/3

1 The student is careful not to use language that indicates certainty, that is, the hypothesis is supported but not proven.

1 The student should be more careful with word choice. They do not increase at the same rate.

The student takes the time to write out a range of comments regarding the hypothesis.

The graph shows a possible (1) inverse correlation ✓ between size and sea temperature as size appears to increase at the same rate (1) as sea temperature drops. At times of higher temperature, size is smallest and at times of relative stability in temperatures there is little change in body size. However, these changes in size could simply be a consequence of the course of evolution not temperature change. Further, the temperatures are estimates and these estimates may be unreliable for the fossil forms. ✓ There is likely very little evidence from older species due to lack of fossils. ✓

Genetic engineering allows genes for resistance to pest organisms to be inserted into various crop plants. Bacteria such as *Bacillus thuringiensis* (Bt) produce proteins that are highly toxic to specific pests.

Stem borers are insects that cause damage to maize crops. In Kenya, a study was carried out to see which types of Bt genes and their protein products would be most efficient against three species of stem borer. The stem borers were allowed to feed on nine types of maize (A–I), modified with Bt genes. The graph below shows the leaf areas damaged by the stem borers after feeding on maize leaves for five days.

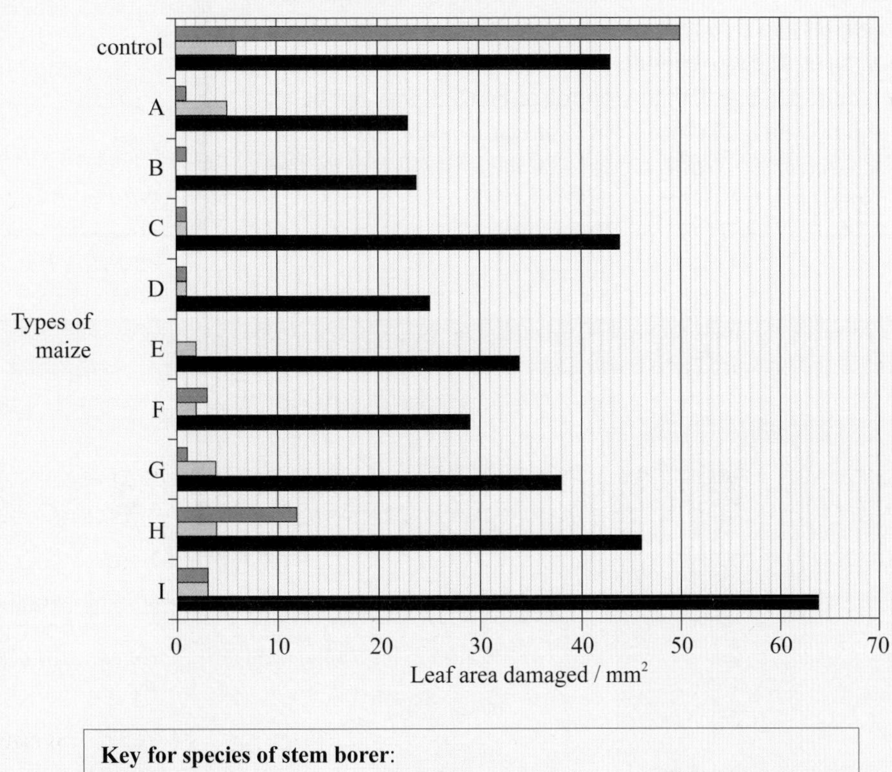

Key for species of stem borer:
■ *Sesamia calamistis* □ *Eldana saccharina* ■ *Busseola fusca*

[*Source*: Mugo, S et al. 2005. *African Journal of Biotechnology*. Vol 4, number 13. Pp 1490–1504. Fig. 3.]

(a) Calculate the percentage difference in leaf area damaged by *Sesamia calamistis* between the control and maize type H. Show your working. *[2]*

(b) Outline the effects of the three species of stem borer on Bt maize type A. *[2]*

(c) Evaluate the efficiency of the types of Bt maize studied, in controlling the three species of stem borers. *[2]*

[Taken from standard level paper 2, time zone 1, May 2009]

How do I approach the question?

(a) Make sure that you are using the correct values: dark grey, control and type H. The background grid helps you find the values in this question, but in other cases you may have needed a ruler. First, calculate the difference (50 – 12); the difference is in relation to the control, so you must divide by the control value (50); this will cancel units, but you must multiply by 100 and indicate "%" since a percentage is asked for.

(b) The question asks you to "outline", so you have to present the general trend, not specific values. Which causes the most damage? Are they causing the same damage to all types?

(c) In this question, you must look at the types of Bt maize studied and write a judgment about their efficiency. The importance and the significance of the data should be part of your answer. You should use the control type as a reference. *S. calamistis* and *E. saccharina* are controlled to some degree by all the Bt maize types. The worse results are with *B. fusca*, where types C, H and I have even more damage than the control, although it may not be significant for C and H. Type B totally controls *E. saccharina*, whereas this is done by type E for *S. calamistis*. Type H seems to be the least efficient for the three stem borer species.

Which areas of the syllabus is this question taken from?

- The mathematical requirements and aim 4 of the syllabus

This answer achieved 1/6

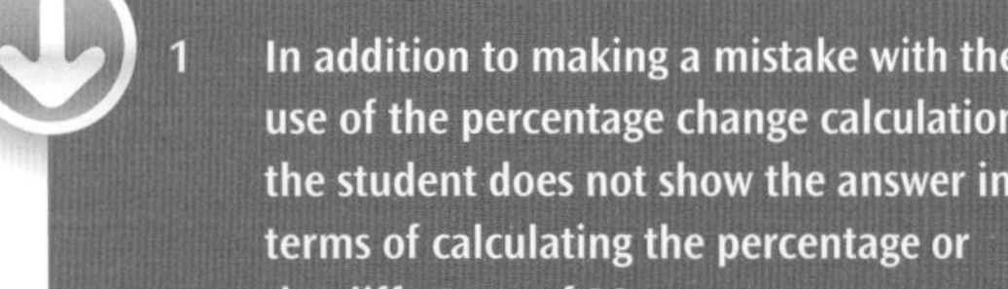

1. In addition to making a mistake with the use of the percentage change calculation, the student does not show the answer in terms of calculating the percentage or the difference of 38.
2. The student has failed to take into account what is occurring in the control group.
3. The student has not summarized the complex data in sufficient detail.

(a) $\frac{50-12}{12} = 316\%$ ❶

(b) Type A maize lowers consumption by the stem borers. ✓

(c) *Eldana saccharina* was the most successfully controlled because it resulted in the least amount of leaf area damaged over all nine types of maize. ❷❸

This answer achieved 3/6

1 The student has managed to answer to the command term "outline" by writing which stem borer caused the most leaf damage and which caused the least.

1 The calculation is incomplete because it only shows the damage for type H compared to the control; the difference with the control is not calculated.

2 The data are not about the stem borers adapting to the maize, but about the maize controlling the stem borers.

3 The student did not relate to any other aspect of the data (for example, all types of Bt maize decrease *Sesamia* damage/*Eladana* damage). The number of marks in brackets is an indication of the number of expected elements in the answer.

(a) $\frac{12}{50} = 24\%$ ❶

(b) *Busseola fusca* showed the most resistance to the Bt in maize type A and caused the most leaf damage ✔ (23 mm^2). *Sesamia calamistis* showed the weakest resistance and did relatively little damage ✔ (1 mm^2). ❶ *Eldana saccharina* had more resistance than *Sesamia* but was unable to do much leaf damage (5 mm^2).

(c) The different maizes all did relatively well in controlling two of the three stem borers, but *Buseola fusca* still managed to adapt ❷ particularly well to the changes in the maize. ✔ ❸

This answer achieved 6/6

1 Even though the question is only worth 2 marks, the student writes out many things noticed in the data.

1 The student's answer is too brief and general. There are more things the student could have pointed out in the data.

Unless the question specifies the number of things to say, it is a good strategy to point out as many relevant aspects of the data that you can, as the student has done in part (c) of this answer.

(a) $\frac{50 - 12}{50}$ ✔ $\times 100\% = 76\%$ ✔

(b) Bt maize variety A was eaten less than the control by all three species of stem borer though the reduction in consumption is very slight for *E. saccharina*. ✔ It reduced consumption by *S. calamistis* the most ✔ and consumption by *B. fusca* is still quite high. ❶

(c) In general, genetic modification of maize reduced consumption by the stem borers. All types of maize reduced consumption by *S. calamistis*. Though most types of maize reduced consumption by *B. fusca*, in some varieties of maize such as type H, ✔ consumption was not reduced but actually went up. All types of maize reduced consumption by *E. saccharina*. ✔ ❶

7. Molecules, cells and metabolism

Cell theory

Useful terminology for this section:

- **cell differentiation**—the process by which certain genes in the genome of multicellular organisms are expressed, causing cells to become specialized in structure and function.

You should know:

- the properties of life
- the principles of cell theory
- emergent properties
- properties of stem cells.

You should be able to:

- list metabolism, response, homeostasis, growth, reproduction and nutrition as properties of life
- outline the cell theory, that living organisms are composed of cells, that cells are the smallest unit of life, and that cells come from pre-existing cells
- discuss the evidence for the cell theory
- compare the relative sizes of molecules, cell membrane thickness, viruses, bacteria, organelles and cells, using the appropriate SI unit
- calculate the linear magnification of drawings and the actual size of specimens in images of known magnification
- explain the importance of the surface area to volume ratio as a factor limiting cell size
- state that multicellular organisms show emergent properties, which arise from the interaction of the component parts—the whole is greater than the sum of its parts
- explain that cells in multicellular organisms differentiate to carry out specialized functions by expressing some of their genes but not others
- state that stem cells retain the capacity to divide and have the ability to differentiate along different pathways
- outline the therapeutic use of stem cells.

Examples

1. Discuss the evidence for the cell theory.

   In this case, "discuss" implies pointing out the facts that back up the cell theory, but also exceptions, including weighing the arguments. A statement of the cell theory is expected. Evidence for it would be that most organisms or parts of organisms are composed of "typical" cells when seen under the microscope. Exceptions would be skeletal cells and fungal hyphae with many nuclei. Unicellular organisms can be considered acellular, but they carry out all life functions. The presence of extracellular material—such as the vitreous humour of the eye, mineral deposits in bone, and cellulose in cell walls—contradicts the idea that the organism is made up of cells only, but this material has been produced by cells.

2. Explain that cells in multicellular organisms differentiate to carry out specialized functions by expressing some of their genes but not others.

   The first principle that should be stated is that all cells in the same organism possess the same genome, that is, the same genes, and could develop in any way. Differentiation could be defined as the development of different or specific structures and functions. Examples of differentiated cells in a multicellular organism could be given. It could be stated that position, hormones, cell-to-cell signals and chemicals determine how a cell develops by switching on or off some genes.

Be prepared

- When multicellular organisms differentiate, they do not express the same genes in all cells, although all cells possess the same genome (chapter 8).
- Stem cells retain the capacity to divide and have the ability to develop along different pathways because they can activate different parts of their genome. That is why they can be used for therapeutic purposes.
- These two principles can be used to elaborate answers to eventual questions about differentiation and stem cells.

Prokaryotic and eukaryotic cells

Useful terminology for this section:

- **fungi** (singular: fungus)—a kingdom in the classification of organisms (chapter 10) containing microscopic organisms. Most fungi are decomposers. Some fungal cells are multinucleate and therefore acellular
- **virus**—a non-cellular structure made of protein and containing a nucleic acid.

You should know:

- structures and functions of prokaryotic and eukaryotic cell components
- structural similarities and differences between animal and plant cells.

You should be able to:

- use the bacterium *Escherichia coli* (*E. coli*) as the required example of a prokaryotic cell
- use a liver cell as the required example of a eukaryotic cell
- draw, label, annotate and identify structures and extracellular components in diagrams and electron micrographs of prokaryotic and eukaryotic cells
- compare prokaryotic and eukaryotic cells
- compare animal and plant cells.

Examples

1. This question refers to the following micrograph of an *E. coli* bacterium undergoing reproduction.

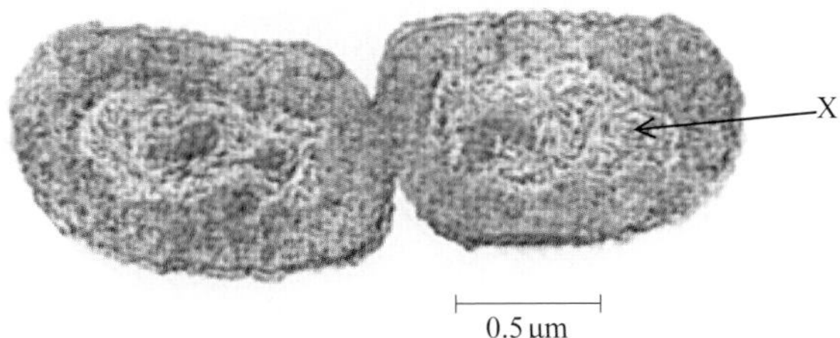

[*Source*: www.bio.mtu.edu/campbell/prokaryo.htm. Copyright © Bedford, Freeman & Worth Publishing Group, LLC.]

In the diagram, what does label X identify?

A. nucleoid region
B. chromatin
C. histones
D. endoplasmic reticulum

This question was written in the hope that the correct answer could be found only by looking at structure X, without having to eliminate possible choices. The arrow points to an area with a different texture: this is the nucleoid region. To proceed, it should be known that *E. coli* is a prokaryote. Only eukaryotes have membrane-bound organelles, like the endoplasmic reticulum, and histones associated with their DNA, so choices C and D are not considered. Chromatin is related to the linear chromosomes of eukaryotes, but bacteria have circular DNA located in the nucleoid region. Knowledge of histones is not required to answer this question.

2. The electron micrograph below shows a section of a liver cell.

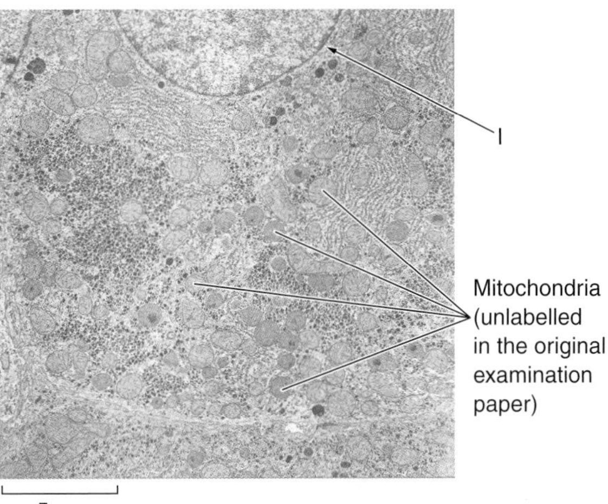

[*Source*: Kalderon, D, Roberts, BL, Richardson, WD and Smith, AE. 1984. "A short amino acid sequence able to specify nuclear location". *Cell*. Vol 39. P 499. Copyright © Elsevier. Reproduced with permission.]

(a) Identify the structure labelled I.
(b) Explain the evidence from the electron micrograph that indicates that liver cells are very active.

The liver cell is the chosen example of a eukaryotic cell in the syllabus. You should be able to identify any structure from the following and know their function: free ribosomes, rough endoplasmic reticulum (RER), lysosome, Golgi apparatus, mitochondrion and nucleus. Here, the nucleus is indicated by arrow I.

Since you should be able to identify mitochondria and know their function (aerobic cell respiration), the abundance of mitochondria (only a few are labelled) on the picture is the sign that liver cells are very active.

Prokaryotic and eukaryotic cells (continued)

Be prepared

- Some questions may ask for comparisons of prokaryotic and eukaryotic cells or between plant and animal cells. The comparisons may include extracellular components.
- Extracellular components refer to the cell wall of plant cells, mainly made out of cellulose, and the glycoprotein extracellular matrix of animal cells.
- The plasma membrane, common to all cells, should not be confused with the cell wall present in bacteria, fungi, plants and a few other organisms. Bacteria are prokaryotic cells.
- Fungi, plants and animals have eukaryotic cells.
- Although viruses are not considered to be living organisms, you are supposed to appreciate the relative size of viruses compared to components of living organisms, know that they may be pathogens (chapter 12), that they can be used in vaccines (chapter 12) and biotechnology (chapter 8) and know about HIV (chapter 12).
- Comparisons between prokaryotic and eukaryotic cells should include the absence/presence of the nucleus, the absence/presence of membrane-bound organelles and the structural features of ribosomes and DNA.
- Some differences between animal and plant cells include a cell wall, chloroplasts and a large (sap) central vacuole in plant cells and centrioles in animal cells, whereas similarities include the presence of a plasma membrane, nucleus, ribosomes and mitochondria.
- Elements in this section may link to further knowledge about cell walls of bacteria for students preparing for option F.

Membranes

Required definitions for this section:

- **diffusion**—the passive movement of particles from a region of high concentration to a region of low concentration
- **osmosis**—the passive movement of water molecules, across a partially permeable membrane, from a region of lower solute concentration to a region of higher solute concentration.

You should know:

- the structure of cell membranes and the properties of their components
- the function of membrane proteins
- the role of membranes in the transport of substances.

You should be able to:

- draw and label the molecular structure of membranes
- explain the hydrophobic and hydrophilic properties of membrane structure phospholipids
- state the definitions of diffusion and osmosis
- explain the mechanisms of diffusion, facilitated diffusion and active transport
- explain the use of transport vesicles within a cell
- describe the change of shape of the membrane in endocytosis and exocytosis.

Examples

1. Draw a labelled diagram to show the fluid mosaic structure of a plasma membrane, indicating the hydrophilic and hydrophobic regions. *[5]*

   As with all drawings, structures should be clearly drawn, correctly labelled and proportional to one another. One mark is usually allocated for each correctly labelled structure; however, sometimes 1 mark is allocated for more than one structure, so it is a good strategy to draw more than five structures, if possible. The phospholipid bilayer should be drawn and labelled, with hydrophilic phosphate heads and two hydrophobic hydrocarbon tails per molecule, tails of the two layers opposed to each other on the inside and heads on the outside. Since the question specifies "plasma membrane" and not only "membrane", proteins should be added, with as many types and with as many details as possible (integral proteins shown spanning the membrane, peripheral proteins shown on the membrane surface and protein channels shown with a channel). Glycoproteins should be shown on the outside and with a carbohydrate side chain; cholesterol molecules in between phospholipids in the hydrophobic region. An indication of thickness (10 nm) would complete the diagram.

Membranes (continued)

2. Explain how the hydrophilic and hydrophobic properties of phospholipids help to maintain the structure of cell membranes. *[6]*

The answer can contain the definition of hydrophilic and hydrophobic and show the structure of a phospholipid molecule with phosphate heads, which are hydrophilic, and two hydrocarbon tails, which are hydrophobic, possibly with a well-labelled diagram. The position of these molecules in relation to each other and water surrounding the cell membrane can then be outlined, with reasons why they are maintained in these positions (hydrophilic molecules are attracted to water, whereas hydrophobic molecules are attracted to one another and repel water). Other statements can complete the answer, such as: phospholipid bilayer or membranes self-assemble in water; protein association with membrane is determined by hydrophobic interactions; and the phospholipid bilayer is hydrophilic on the outside and hydrophobic on the inside.

Be prepared

- The term "membrane" may refer to the plasma membrane or the inner foldings of a eukaryotic cell forming specific organelles. Both are formed by a phospholipid bilayer, but the rest of their composition may vary.
- Knowledge of the difference between integral proteins embedded in the phospholipids of the membrane and of peripheral proteins attached to the membrane surface is required.
- A distinction between the different processes of transport across membranes, their usage and their requirement of a supply of energy is required.
- Transport of substances in vesicles can be linked to other processes covered by the syllabus, for example transcription in protein synthesis (chapter 8).

Cell division and mitosis

You should know:

- processes in the cell cycle
- when binary fission and mitosis are involved
- the events that happen in mitosis.

You should be able to:

- state that the stages in the cell cycle include interphase and cell division (mitosis and cytokinesis)
- describe the events that occur in the four stages of mitosis
- state that growth, embryonic development, tissue repair and asexual reproduction involve mitosis
- explain how mitosis produces two genetically identical nuclei
- relate protein synthesis (chapter 8), DNA replication (chapter 8) and increase in mitochondria and chloroplasts to interphase.

Examples

1. By what process do most bacteria divide?
   A. mitosis
   B. meiosis
   C. conjugation
   D. binary fission

The correct answer should be recalled as factual information about bacteria, but can also be deduced. Mitosis and meiosis involve linear chromosomes, which are absent in bacteria (prokaryotic cells). The purpose of mitosis is to make two identical eukaryotic cells, whereas the purpose of meiosis is to prepare gametes (reproductive cells) by reducing by half the number of chromosomes of eukaryotic cells. Thus, choices A and B are incorrect. Conjugation is the process by which bacteria exchange genetic material; the bacteria may divide afterwards, but conjugation is not cell division. Binary fission (D) is therefore the only choice left.

2. What is a difference between a cell in the G_1 phase and a cell in the G_2 phase in the cell cycle.
   A. A cell in the G_2 phase would be smaller than a cell in the G_1 phase.
   B. A cell in the G_2 phase would have more mitochondria than a cell in the G_1 phase.
   C. A cell in the G_1 phase would have more DNA in its chromosomes than a cell in the G_2 phase.
   D. DNA replication occurs in the G_1 phase but not in the G_2 phase.

The question assumes knowledge of the names of the cell cycle phases, the ability to outline the process(es) taking place in each part of interphase and knowledge of the sequence of the stages. Knowledge of DNA replication in the second phase, the S phase, rules out choices C and D. A cell in the later G_2 phase would be larger than the earlier G_1 phase and would have more mitochondria, leaving B as the right answer.

Cell division and mitosis (continued)

3. Explain how mitosis produces two genetically identical nuclei.

   Mitosis occurs after DNA replication. Therefore the genetic material in the nucleus has already doubled before the onset of mitosis. The two DNA molecules resulting from DNA replication in each chromosome coil separately to form the two sister chromatids of each chromosome. Sister chromatids (of a chromosome) are therefore identical. Chromosomes line up on the same plane during metaphase. Chromatids of each chromosome split and move in opposite directions during anaphase and form the chromosomes in the nucleus of each of the two new cells. Since each chromatid was identical to its sister, the resulting nuclei are identical.

Be prepared

- Your knowledge that growth, embryonic development, tissue repair and asexual reproduction involve mitosis should be applied to the specifics of exam questions. For instance, humans do not reproduce asexually, and unicellular organisms do not go through embryonic development.
- You should also be able to relate the stages of the cell cycle to metabolic processes such as DNA replication, transcription, translation (all in chapter 8), increase in the number of mitochondria and/or chloroplasts.
- Many new discoveries about cancer are made all the time. Although tumours are the result of uncontrolled cell division, gene mutation, chromosome mutation (chapter 8) and gene expression are involved in various ways. Data from cancer research may be used for analysis in exam papers.

Chemistry of life

Useful terminology for this section:

- **inorganic compounds**—compounds that do not contain carbon or contain only one atom of carbon (such as hydrogen carbonates, carbonates and oxides of carbon)
- **organic compounds**—compounds containing carbon that are found in (or made by) living organisms (except hydrogen carbonates, carbonates and oxides of carbon).

You should know:

- structure and properties of water
- most common elements in organisms
- categories and functions of organic compounds found in organisms.

You should be able to:

- draw the structure of water molecules, showing polarity and hydrogen bonding
- explain the relationship between the thermal, cohesive and solvent properties of water and its use as a coolant, a medium for metabolic reactions and transport in organisms
- state that carbon, oxygen, hydrogen and nitrogen are the most common elements in organisms
- state the roles of sulfur, calcium, phosphorus, iron and sodium in prokaryotes, plants and animals
- list glucose, fructose and galactose as monosaccharides, maltose, lactose and sucrose as disaccharides, and starch, glycogen and cellulose as polysaccharides
- state one function of glucose, lactose and glycogen in animals, one function of fructose, sucrose and cellulose in plants, and three functions of lipids
- outline how carbohydrates, lipids and amino acids or polypeptides combine through condensation and split through hydrolysis in organisms
- identify amino acids, glucose, ribose and fatty acids from diagrams showing their structure
- compare the use of carbohydrates and lipids in energy storage.

Example

Blood is a water-based medium. Which property of water makes it a good transport medium.

A. high specific heat
B. transparency
C. versatility as a solvent
D. it has its greatest density at 4°C

C is the correct answer. The question statement links blood to the three water properties to be known; it asks you to link them to the transport of substances. Only the solvent property (C) enables water, and therefore blood, to transport many substances.

Chemistry of life (continued)

Be prepared

- You should know the names of the required organic molecules in the syllabus, and be able to recognize their structure. Understanding their functions in different types of organisms is a must, as this knowledge can be linked to other parts of the syllabus within IB exam questions.
- Many molecules, including water, show polarity or non-polarity. Their behaviour when they are in the presence of other molecules can be explained by interactions brought about by polarity, as well as by bonds between atoms, including hydrogen and sulfur.
- Use the terms "specific heat" and "heat of vaporization" in the proper context when explaining water's significance for life. Both are thermal properties, but with quite different meanings.

Enzymes

Required definitions for this section:

- **active site**—surface location on an enzyme where it binds and interacts with the substrate
- **denaturation**—structural change in an enzyme (caused by heat above optimum temperature or by pH above or below the optimum pH) that destroys the enzyme's functionality
- **enzyme**—a globular protein that can catalyse a reaction and cause structural change in another molecule (the substrate) by lowering the activation energy of the reaction.

You should know:

- mechanisms of enzyme activity
- use of enzymes in biotechnology.

You should be able to:

- relate enzymes to globular proteins
- explain specificity between enzyme and substrate, with reference to the lock-and-key model
- explain how temperature, pH and substrate concentration affect enzyme activity
- explain the use of lactase in the production of lactose-free milk.

Example

Explain the effect of pH on enzyme-catalysed reactions. *[3]*

Statements that enzymes have a pH optimum and that the active site works best at this pH can start the answer. Then it can be shown that activity decreases above and below the optimum pH, perhaps using a labelled diagram. The mechanism of interfering with hydrogen bonding and active site structure should be part of an explanation. The answer can be completed by stating that extremes of pH denature the active site or enzyme, so enzyme activity or reaction stops.

Be prepared

- The influence of factors on enzyme activity can be questioned by data or graphs provided in exam papers.
- The syllabus is limited to proteins acting as enzymes, but research has demonstrated that other molecules, such as RNA (chapter 8), can exhibit enzymatic properties. You may be asked to apply your knowledge about enzymes and RNA to new situations involving such molecules and suggest outcomes based on data provided.
- For a question asking to explain a reason for converting lactose to glucose and galactose during food processing, just writing "some people are lactose intolerant" would be incomplete. "Explain" (giving a detailed account of causes, reasons or mechanisms) would require an expanded answer. For example, "some people are lactose intolerant or have difficulty digesting lactose but could consume milk products where lactose has been converted to glucose and galactose" would be complete.

Cell respiration

Useful terminology for this section:

- **cell respiration**—the controlled release of energy from organic compounds in cells to form ATP.

You should know:

- the outcomes of cell respiration
- the principles of anaerobic cell respiration
- the principles of aerobic cell respiration.

You should be able to:

- outline glycolysis, including its location and products of glucose breakdown
- explain anaerobic and aerobic cell respiration
- state what happens to pyruvate when conditions are anaerobic or when conditions are aerobic.

Examples

1. Distinguish between the process of anaerobic respiration in yeast and humans. *[2]*

yeast: pyruvate to ethanol and carbon dioxide

humans: pyruvate to lactic acid

Pyruvate had to be mentioned to gain full marks if end-products were correct.

2. Which equation shows a chemical reaction that occurs during anaerobic cell respiration?

A. pyruvate → lactate

B. pyruvate → lactate (ADP → ATP)

C. pyruvate → lactate + CO_2

D. pyruvate → ethanol + ATP

Pyruvate can be converted into lactate, or ethanol and CO_2 during anaerobic cell respiration. This happens with no further yield of ATP, and therefore choices B and D are incorrect answers. Choice D is also incorrect for a second reason because no CO_2 is shown. Choice C is incorrect since production of lactate does not release CO_2. Only choice A represents a correct answer.

Be prepared

- Fermentation usually refers to anaerobic respiration by microorganisms, producing ethanol and CO_2, whereas anaerobic respiration in humans produces lactate and does not release CO_2.

Photosynthesis

Useful terminology for this section:

- **photolysis**—the process by which light energy splits water molecules into hydrogen and oxygen in chloroplasts.

You should know:

- the role of sunlight and pigments in photosynthesis
- the principles of the production of ATP and organic molecules in photosynthesis
- how to measure the rate of photosynthesis
- the influence of various factors on the rate of photosynthesis.

You should be able to:

- state that photosynthesis involves the conversion of light energy into chemical energy
- state that sunlight is composed of a range of wavelengths corresponding to colours and chlorophyll absorbs mainly red and blue wavelengths but reflects the green
- state that chlorophyll is the main photosynthetic pigment
- state that light energy is used to split water molecules (photolysis) into hydrogen and oxygen
- state that ATP and hydrogen (from photolysis) are used to fix carbon dioxide to make organic molecules
- explain that the rate of photosynthesis can be measured directly by the production of oxygen or uptake of carbon dioxide, or indirectly by an increase in biomass
- outline the effects of temperature, light intensity and carbon dioxide concentration on the rate of photosynthesis.

Examples

1. What reaction, involving glycerate 3-phosphate, is part of the light-independent reactions of photosynthesis?

   A. Glycerate 3-phosphate is carboxylated using carbon dioxide.

   B. Two glycerate 3-phosphates are linked together to form one hexose phosphate.

   C. Glycerate 3-phosphate is reduced to triose phosphate.

   D. Five glycerate 3-phosphates are converted to three ribulose 5-phosphates.

   This question focuses on the recall of factual information about the light-independent reactions of photosynthesis (Calvin cycle). Molecule names may vary according to textbooks, but only names mentioned in the IB Diploma Programme biology standard level syllabus are used here. Only glycerate 3-phosphate is reduced to triose phosphate, so C is the correct answer.

2. Outline how three different environmental conditions can affect the rate of photosynthesis in plants. *[6]*

   Only three factors are mentioned in the syllabus: temperature, light intensity and carbon dioxide concentration. Expect that marks could be split equally between these three. Answers can use diagrams, provided that they are annotated clearly.

   The rate of photosynthesis increases with increasing light intensity, but it reaches a plateau. Rate increases with increasing carbon dioxide level to a maximum, then plateaus as photosynthesis reaches an optimal level.

Be prepared

- Absorption and reflection of wavelengths may be represented by graphs in exams.
- You should be able to explain the shapes of graphs that show the effect of temperature, light intensity and carbon dioxide concentration on the rate of photosynthesis.
- Explaining the process of photosynthesis requires a detailed account of events, in a logical or chronological order and using appropriate terminology.
- For light-dependent and light-independent reactions, students should understand that products of one reaction become the substrates of the other and vice versa, and that this can become a reason for a plateau in the rate of reaction but that light is not renewable.

Explain the importance of the surface area to volume ratio as a factor limiting cell size. *[3]*

How do I approach the question?

- A relatively short answer is expected.
- There are 3 marks, and therefore the answer should contain three or four important elements.
- The command term is "explain"; causes and/or mechanisms are expected.
- The word "importance" suggests that the surface area to volume ratio should be treated as a cause.
- Some principles should be stated to build up the explanation. The rate of exchange between a cell and its environment is a function of its surface area. The membrane is used to exchange nutrients, gases, products, heat and waste substances. Cell activities/metabolism are a function of the cell's volume. Heat, products and waste are the result of metabolism.
- It should be stated that surface area increases by the square, whereas volume increases by the cube. Therefore, volume increases much faster than surface area. This can be demonstrated with an example such as a cube using a small table with three lines comparing side, surface area, volume and surface area to volume ratio.
- The argument that an increase in cell volume increasingly reduces the cell's exchange capability to a point where metabolism is limited is the turning point of the answer, followed by the argument that dividing the cell restores the original ratio as a larger surface area is available for exchanges.

Which areas of the syllabus is this question taken from?

- Importance of surface area to volume ratio (2.1.6)

This answer achieved 0/3

In this response, the student starts by restating the question. The student should expand the explanation.

Surface area to volume ratio is important because it limits the cell size. It helps the cell support itself. It helps to contain heat, water and lastly oxygen.

This answer achieved 2/3

There is not sufficient detail around the mechanism (production of waste and heat is a function of volume). An explanation usually requires identification of reasons and/or mechanisms (the "why" and the "how"), and for this question both were expected.

The student writes clearly and provides reasons.

The larger the cell becomes, the smaller the surface area to volume ratio becomes. ✔ This is an issue because cells need to be able to absorb resources and to expel wastes and heat produced. With a small surface area to volume ratio, cells don't have the necessary surface area to perform such actions, ✔ thus creating a cell that lacks resources and is unable to rid itself of toxins.

This answer achieved 3/3

1. A valid mathematical argument is presented.
2. Biological arguments are presented for both surface and volume.

The more a structure increases the surface area increases by square but the volume increases by cube. 1 ✔

Function of volume: waste production, heat production, resource consumption. ✔

Function of surface area: bringing in resources. ✔ 2

Draw a labelled diagram showing the ultra-structure of a liver cell. *[4]*

[Taken from higher level paper 2, time zone 2, May 2009]

How do I approach the question?

- There are 4 marks, and therefore at least four structures must be correct. "Draw" means represent by means of pencil lines. Ultra-structure is about what can be seen with an electron microscope, so details of the structure of organelles must be present.
- It must represent a liver cell. Labels must be added—they should point **on** the structures.
- Appropriate terms must be used, with required distinctions (for example, "**rough** endoplasmic reticulum" instead of "endoplasmic reticulum" only).

Which areas of the syllabus is this question taken from?

- Knowledge of what should be represented in a liver cell (2.3.1)
- Relative size of [...] organelles and cells (2.1.4)
- Distinction between prokaryotic and eukaryotic cells (2.2 versus 2.3)
- Differences between plant and animal cells (2.3.5)

This answer achieved 1/4

1. A double membrane with pores is expected (see other examples).
2. The student has pointed to a nucleolus when referring to the nucleus.
3. Mitochondria are too large in comparison to the nucleus.
4. Proportions are important. Here, the plasma membrane is too thick compared to the shown organelles. The double phospholipid layer can only be seen using a much higher magnification. The expectation was that the plasma membrane would be represented by a single line.

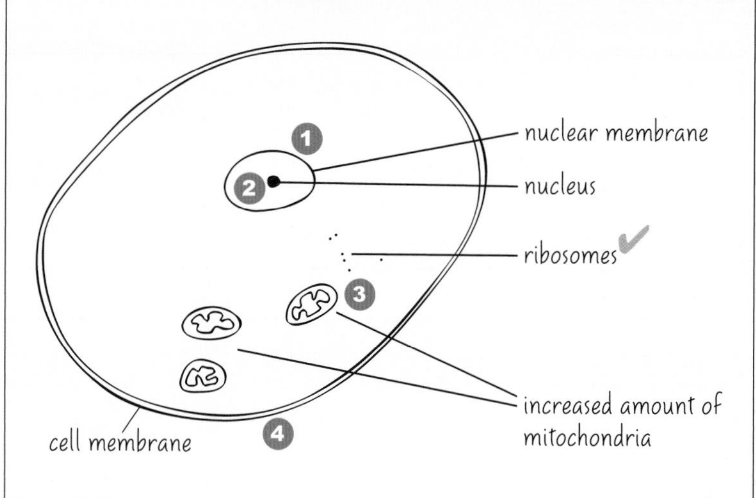

There are not enough elements labelled to cover 4 marks. The number of marks is an indication of the number of elements required in the answer.

This answer achieved 4/4

Despite the drawing being more minimalistic than the next one, there are enough elements to gain full marks. The student has shown the microvilli typical of a liver cell. Drawing and labelling more than four features (as suggested by the number of marks allocated for this question) paid off, and the student gained full marks.

1 It is good practice to include a title.

1 Mitochondria are too large in comparison to nucleus. A double membrane, with folds of the inner membrane is expected. The mitochondria should be drawn with a double line, representing the outer and inner membranes. The cristae should be represented as folds of the inner membrane.

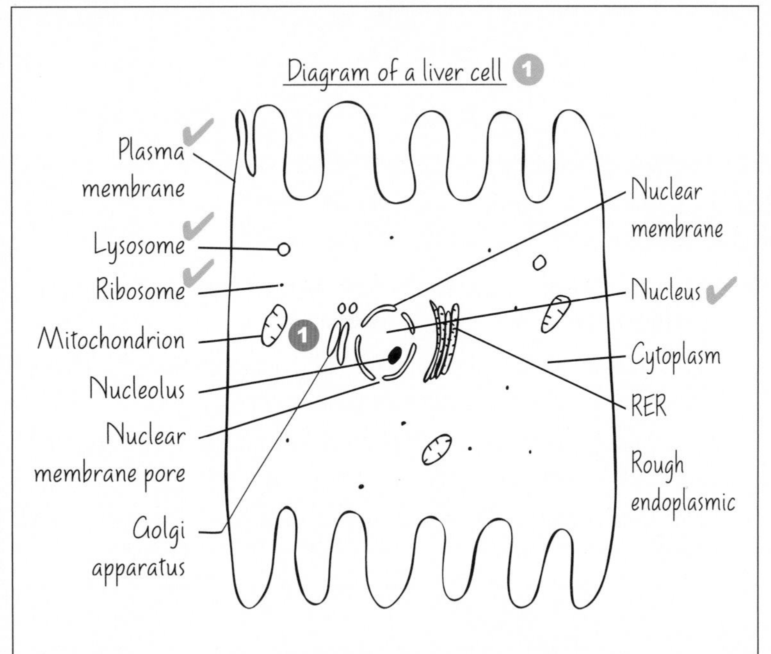

This answer achieved 4/4

1 The student took care drawing nuclear envelope pores.

2 The student has shown that there could be numerous organelles of the same type within a cell, although marks are usually granted for one structure clearly drawn and labelled.

1 This should show folds (cisternae), but since they are shown for the smooth endoplasmic reticulum, the examiner gave the mark. Also rough endoplasmic reticulum should be closer to the nucleus.

This is not a typical liver cell as mentioned in the Diploma Programme *Biology guide* (March 2007), but there are enough positive elements to gain full marks.

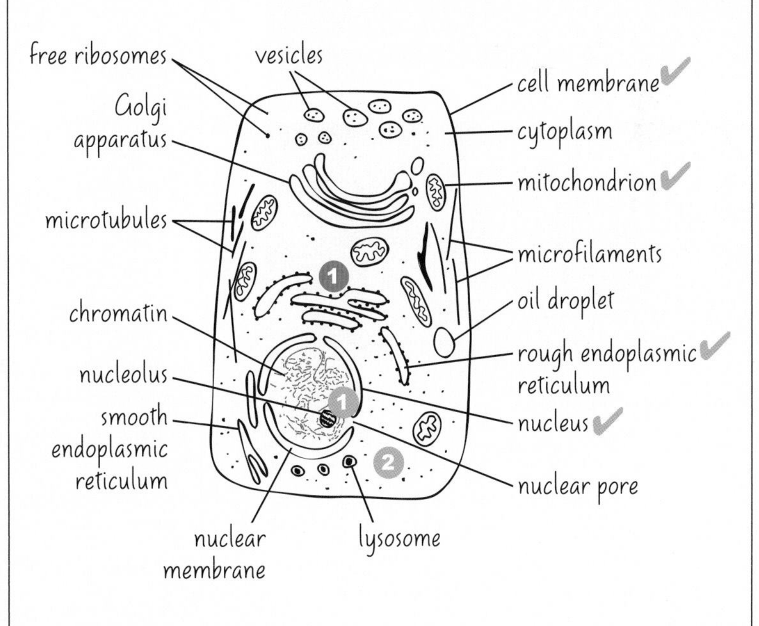

Overall, a very neat diagram, well labelled, with a good attempt to represent shapes, proportions and details.

Distinguish between prokaryotic cells and eukaryotic cells. *[6]*

[Taken from higher level paper 2, time zone 2, May 2009]

How do I approach the question?

- "Distinguish" means give the difference between two or more items. It is therefore not necessary to show similarities.
- It is likely that marks will be allocated for pairs of comparable elements. A one-to-one approach is therefore a good strategy.
- Presenting the distinctions within a two-column table, with opposed statements on each line, prevents mistakes and saves time.
- There are 6 marks, and therefore at least six pairs of statements must be correct.

Which areas of the syllabus is this question taken from?

- Comparison of prokaryotic and eukaryotic cells (2.3.4)

This answer achieved 2/6

1. The student has drawn a table, which is a good start in dealing with this question.
2. A few comparable elements are present.

1. **Unrelated elements are opposed, showing a poor understanding of structures.**
2. **Incorrect facts have been included.**
3. **There is confusion between areas of the syllabus, here plant and animal cells.**

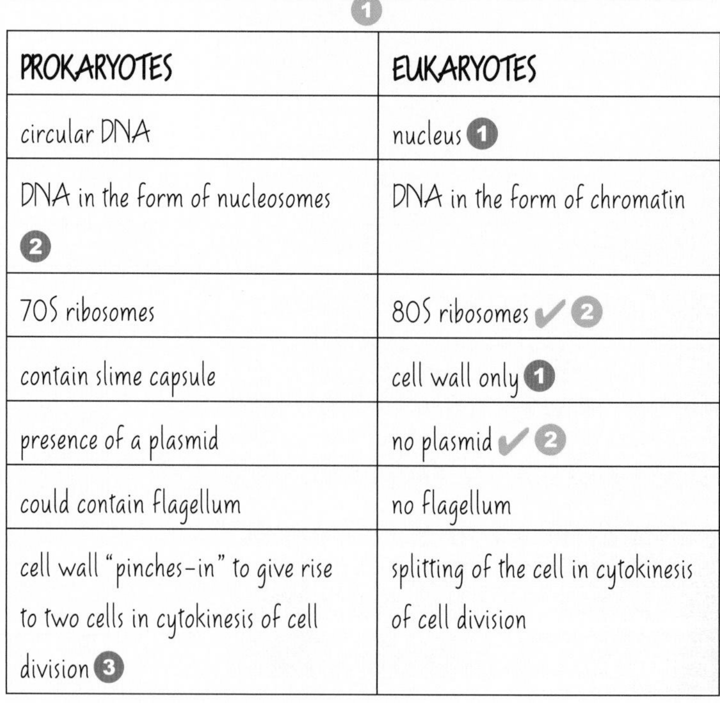

❶

PROKARYOTES	EUKARYOTES
circular DNA	nucleus ❶
DNA in the form of nucleosomes ❷	DNA in the form of chromatin
70S ribosomes	80S ribosomes ✓ ❷
contain slime capsule	cell wall only ❶
presence of a plasmid	no plasmid ✓ ❷
could contain flagellum	no flagellum
cell wall "pinches-in" to give rise to two cells in cytokinesis of cell division ❸	splitting of the cell in cytokinesis of cell division

This answer achieved 3/6

1. The student has drawn a table, which is a very effective way to deal with this question.
2. The student opposes elements on each line.

1. The student has listed organelles instead of the principle of membrane-bound organelles.
2. Unclear statements or statements not using appropriate terminology cannot be awarded a mark. The mention that prokaryotes have rigid flagella whereas eukaryotes have membrane bound flagella was required.
3. Prokaryotes have cell walls, but not all eukaryotes have one, which was not specified.
4. Similarities are not expected with the command term "distinguish".

Prokaryotic cells	❶ Eukaryotic cells
ribosomes are 70S (Svedberg units)	ribosomes are 80S ✔ ❷
has a nucleoid (no nucleus)	has a nucleus ✔
circular DNA	linear DNA ✔
no Golgi	has Golgi ❶
no rough endoplasmic reticulum	has rough endoplasmic reticulum
flagella are unsupported	flagella are supported ❷
no cell wall	cell wall ❸
no lysosomes	has lysosomes

both have a form of DNA ❹
both have a plasma membrane
both have vesicles

This answer achieved 5/6

1. The student opposes distinguishing elements systematically ...
2. ... adding details where available.
3. "Eukaryotic cells have mitochondria whereas prokaryotic cells do not" was the correct and expected statement. Many students in this exam paper have opposed mitochondria to mesosomes instead, and this was incorrect. Mesosomes are now regarded as an artefact of preparation for electron microscopy. The current IB Diploma Programme biology standard level syllabus does not refer to mesosomes.

1. Not enough is given here for the mark. The fact that plants and fungi have cell walls should be mentioned for eukaryotic cells.

This is a well-structured answer, even if a table was not used.

All animals and plants are composed of eukaryotic cells. Bacteria on the other hand are composed of prokaryotic cells. Eukaryotic cells have a nucleus, whereas prokaryotic cells do not, ✔ ❶ prokaryotic cells only have a nucleoid region. Secondly, eukaryotes have their chromosomes (DNA) enclosed in a nucleus, ❷ in addition the chromosomes are supported by contain proteins. Prokaryotic cells on the other hand have naked DNA without proteins. ✔ Furthermore, while prokaryotic cells have cell walls, eukaryotic cells do not, ❶ they only have a plasma membrane & (prokaryotic cells also have a plasma membrane). Then, eukaryotes have membrane bound organelles to compartmentalize functions while prokaryotic cells do not. ✔ For example, eukaryotic cells have mitochondria whereas prokaryotic cells do not. ❸ ✔ Moreover, whereas eukaryotic cells have 80S ribosomes, prokaryotic cells have smaller 70S ribosomes. ✔ Lastly, eukaryotic cells are often bigger than prokaryotic cells. Eukaryotic cells can be up to 10 μm whereas certain prokaryotic cells such as bacteria are only 0.5 μm.

(a) Define active site. [1]

(b) Explain enzyme–substrate specificity. [3]

[Taken from standard level paper 2, time zone 1, May 2009]

How do I approach the question?

(a) This part is based on the recall of factual information, as most definitions are. A definition must be stated using appropriate terminology and incorporating correctly the key ideas, which are "portion of the enzyme", "binding" and "substrate".

(b) This part implies an explanation, although brief, as to why there is specificity. An answer could start with characteristics of both enzyme and substrate, pointing out that they have a three-dimensional shape, giving them their specificity. The mention of fitting shapes (for example, lock and key) is important, but it should not stop there, as many students did for this exam. It should be mentioned that the proximity of structures enables the chemical reaction to take place. It is also worth mentioning that the enzyme is not part of the reaction, it only catalyses it. The answer could be completed by a named example of an enzyme and its substrate, and an annotated diagram.

Which areas of the syllabus is this question taken from?

- Define *enzyme* and *active site* (3.6.1)
- Explain enzyme-substrate specificity (3.6.2)

This answer achieved 1/4

1 There is some understanding that enzyme shape must match substrate shape.

1 The student does not mention substrate.

2 The active site is not on the substrate. Too many students confuse enzyme and substrate.

3 "Specific purpose" is vague. A reference to matching shapes would be better.

(a) An active site is the location on an enzyme where reactions are catalysed. (1)

(b) This means that only certain enzymes can bind to certain active sites on substrates, (2) because each enzyme has a specific purpose. (3) Some enzymes will not bind to the active sites of substrates because they do not fit. ✓ (1)

This answer achieved 4/4

1 "Location" would have been preferable to "space", but the definition is nevertheless complete.

The explanation includes the concept of compatible shapes, uses a valid example, and refers to mechanism of catalysis.

(a) The space on an enzyme (a globular protein which catalyses reactions) where the enzyme's specific substrate(s) join to the enzyme in order for a reaction to be catalysed. ✔ 1

(b) An enzyme's active site is shaped specifically to fit only certain substrates ✔ (often one or two) in order to catalyse a specific reaction. For example, the lactase ✔ enzyme has an active site into which only lactose can fit; lactose is lactase's "specific substrate" which is then broken down into glucose and galactose with lactase's help. ✔

8. Nucleic acids

DNA structure

You should know:

- the basic structure of DNA
- the principles of complementary base pairing.

You should be able to:

- outline the structure of a DNA nucleotide, which is composed of the sugar *deoxyribose* linked to a phosphate and to a base on the other end
- state that the four bases in DNA are adenine (A), thymine (T), cytosine (C) and guanine (G)
- outline and/or draw and label a simple diagram of the molecular structure of DNA showing nucleotides linked by covalent bonds between sugars and phosphates into a single strand and by hydrogen bonds to the other strand
- explain how a DNA double helix is formed using adenine–thymine and cytosine–guanine complementary base pairing and hydrogen bonds.

Example

The diagram below represents part of the DNA molecule.

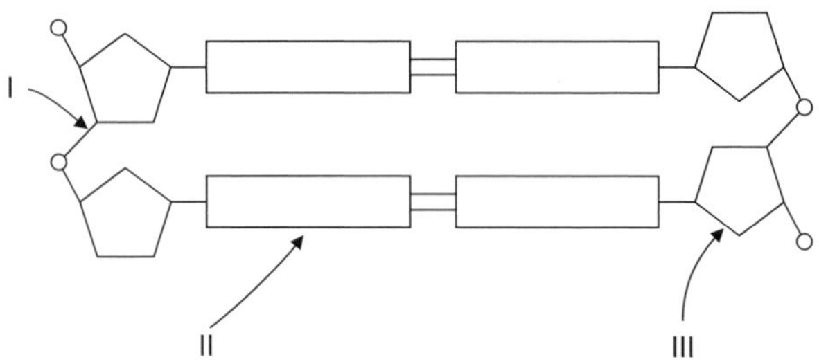

What are the parts labelled I, II and III?

	I	II	III
A.	hydrogen bond	base	deoxyribose
B.	hydrogen bond	deoxyribose	phosphate group
C.	covalent bond	base	deoxyribose
D.	covalent bond	deoxyribose	phosphate group

The syllabus specifies that you should be able to recognize the basic structure of DNA. Commonly used shapes are a circle for the phosphate group, a pentagon for the deoxyribose (this is DNA) and a rectangle for the bases. Phosphates and deoxyriboses are linked with covalent bonds, whereas complementary bases are linked with hydrogen bonds. The correct answer is therefore C.

Be prepared

- Although deoxyribose need only be depicted as a pentagon (without any numbered carbon atoms), labeled covalent bonds should link to the corners of the pentagon figure where carbon atoms exist.

DNA replication

You should know:

- the principles of DNA replication
- how complementary base pairing enables replication to be semi-conservative.

You should be able to:

- explain DNA replication in terms of unwinding the double helix and separation of the strands by helicase, followed by the formation of the new complementary strands by DNA polymerase
- state that DNA replication is semi-conservative
- explain the significance of complementary base pairing in the conservation of the base sequence of DNA.

Examples

1. If 15% of the nucleotides in a sample of DNA contain thymine, what percentage of the nucleotides would contain guanine?
   A. 15%
   B. 30%
   C. 35%
   D. It cannot be determined from the information.

This question is based on the understanding of complementary base pairing. There should be an equal percentage of thymine and adenine, and of cytosine and guanine. Adenine should account for 15%. If A + T = 30%, the rest is 70%, accounted for by C + G. Guanine is half of this amount, 35%. The correct answer is C.

2. What is replicated by a semi-conservative process?
   A. Messenger RNA (mRNA) only
   B. Messenger RNA (mRNA) and transfer RNA (tRNA) only
   C. Messenger RNA (mRNA), transfer RNA (tRNA) and DNA only
   D. DNA only

You should clearly distinguish between replication, transcription and translation. Only DNA is replicated, and it happens that the process is semi-conservative. D is the correct answer and the other choices are all irrelevant.

Be prepared

- It is a good practice to explain DNA replication in the chronological sequence of the events.

Transcription

You should know:

- differences and similarities between RNA and DNA
- the process of transcription.

You should be able to:

- compare the structure of RNA and DNA
- outline the formation of a messenger RNA (mRNA) strand complementary to the DNA strand by RNA polymerase.

Example

Distinguish between RNA and DNA. *[3]*

The command term "distinguish" only requires a statement of differences. Here, three are required since there are 3 marks allocated. The sugar in RNA is ribose, whereas in DNA it is deoxyribose. Uracil is specific to RNA, whereas thymine is specific to DNA. RNA is usually single-stranded, whereas DNA is usually double-stranded. If the question had required a comparison, similarities would also have been required: adenine, cytosine and guanine are common to DNA and RNA, and both uracil and thymine are complementary to adenine.

Translation

You should know:

- the composition of the genetic code
- the general process of translation
- the relationship between one gene and one polypeptide.

You should be able to:

- describe the genetic code as a series of triplets of bases known as *codons*
- explain the general process of translation leading to polypeptide formation, including the roles of mRNA (messenger RNA), tRNA (transfer RNA), codons, anticodons, ribosomes and amino acids
- discuss the relationship between one gene and one polypeptide.

Example

What is a codon?

A. A sequence of nucleotides on rRNA that corresponds to an amino acid
B. A sequence of nucleotides on mRNA that corresponds to an amino acid
C. A sequence of nucleotides on tRNA that corresponds to an amino acid
D. A gene that corresponds to a polypeptide

A codon is a group of three m-RNA nucleotides by definition. B is therefore the correct answer. Triplets of nucleotides on t-RNA are known as anti-codons, thus eliminating choice C. rRNA does not directly transfer sequences of information (and is not addressed by the syllabus anyways). Choice D is irrelevant because the idea of triplets is not involved.

Be prepared

- Anticodons are a part of tRNAs. The amino acids are carried on the tRNAs, and it is inappropriate to say that they are carried by the anticodons.
- An explanation of translation should use appropriate terminology and should reflect the order of the steps involved.

Chromosomes, genes, alleles and mutations

Required definitions for this section:

- **allele**—one specific form of a gene, differing from the other alleles by one or a few nucleotides only and occupying the same gene locus as other alleles of the gene
- **gene**—a heritable factor that controls a specific characteristic
- **gene mutation**—a sudden and irreversible change involving one or many nucleotides of a gene
- **genome**—the whole of the genetic information of an organism.

You should know:

- the composition of eukaryote chromosomes
- the definitions of gene, allele, genome and gene mutation
- the consequence of mutations.

You should be able to:

- state that eukaryote chromosomes are made of DNA and proteins
- define gene, allele, genome and gene mutation
- explain that a base substitution mutation in the DNA will cause the transcription into a different codon in the mRNA, which may result in the insertion of a different amino acid in a polypeptide during transcription.

Examples

1. What is the cause of sickle-cell anemia?

   A. A change to the base sequence of a hemoglobin gene
   B. Mosquitoes acting as the vector for malaria
   C. Iron deficiency due to the malaria parasite
   D. Production of more white blood cells than red blood cells by bone marrow

The cause of sickle-cell anemia as a change to the base sequence of a hemoglobin gene (choice A) is part of the syllabus. (See also pages 45, 49 and 53.)

Chromosomes, genes, alleles and mutations (continued)

2. Which of the following chemicals is a component of eukaryotic chromosomes?
   A. Protein
   B. Triglyceride
   C. Fatty acid
   D. RNA

This question is based on the knowledge required by the syllabus. Besides DNA, only proteins (choice A) are components of eukaryotic chromosomes.

Be prepared

- The consequence of a base substitution mutation must be explained using the example of sickle-cell anemia, where the substitution of GAG by GTG causes the substitution of glutamic acid by valine during the transcription of the hemoglobin protein.
- Gene mutations give rise to different alleles for a gene.
- Theoretical genetics study the combinations in the inheritance of alleles through generations.

Genetic engineering techniques

You should know:

- the principles of the polymerase chain reaction (PCR) technique
- the principles and use of gel electrophoresis.

You should be able to:

- outline the use of the polymerase chain reaction (PCR) to copy and amplify minute amounts of DNA
- state that fragments of DNA are separated according to their size and move in an electric field in gel electrophoresis
- describe the application of DNA profiling to determine paternity and also in forensic investigations
- analyse DNA profiles to draw conclusions about paternity or forensic investigations
- outline three outcomes of the sequencing of the complete human genome.

Examples

1. What conclusions can be made from the following evidence from an analysis of DNA fragments?

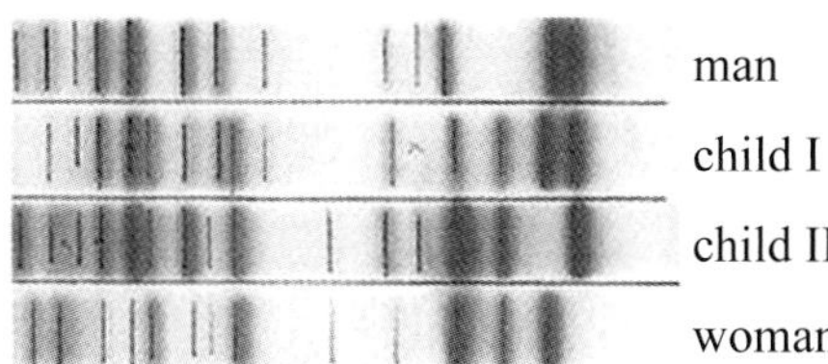

   A. Both children are related to both parents
   B. Child I is related to the man but child II is not.
   C. Both children are unrelated to either of the parents
   D. Child II is related to the man but child I is not.

A relationship between the children and the parents, if it exists, should be demonstrated by the presence of common bands on the electrophoresis pattern. Both children share bands with the two parents, so choice C is excluded. They also share bands with the man, although they may be different. A is therefore the correct answer.

2. What could be achieved by DNA profiling using gel electrophoresis?
   A. The chromosome number of an organism could be counted.
   B. It could be proven that human tissue found at the site of a crime did not come from a person suspected of having committed the crime.
   C. A karyotype could be produced.
   D. Extinct species of living organisms could be brought back to life.

Only a small portion of an individual's DNA is used in gel electrophoresis, and this DNA has been cut into fragments. Karyotypes are made from photos of all chromosomes of a cell. Only B is a possible answer.

Be prepared

- Although the use of restriction enzymes (endonucleases) is associated with gene transfer in the syllabus, restriction enzymes are also used to fragment DNA prior to gel electrophoresis.
- The sizes of DNA fragments in gel electrophoresis are often expressed in multiples of base pairs (bp) (for example, kbp), since they are parts of DNA involving the two strands.
- Graphs and/or illustrations expressing these units may be used in exam papers.

Cloning, and gene transfer and modification

Required definition for this section:

- **clone**—a group of genetically identical organisms or a group of cells derived from a single parent cell.

You should know:

- the universality of the genetic code
- the principles of gene transfer
- examples of genetic modification
- principles of cloning.

You should be able to:

- state that the transfer of genes between organisms will result in polypeptides with the same amino acid sequence because of the universality of the genetic code
- outline a basic technique for gene transfer involving plasmids, a host cell, restriction enzymes (endonucleases) and DNA ligase
- state the current use of genetically modified tomatoes for salt tolerance, rice for beta-carotene and vitamin A, crop plants for herbicide resistance, and sheep milk for human blood clotting factor IX
- discuss the potential benefits and possible harmful effects of one example of genetic modification
- outline a technique for cloning using differentiated animal cells
- discuss the ethical issues of therapeutic cloning in humans.

Example

Using a **named** example, discuss the benefits and harmful effects of genetic modification. *[9]*

From the text of the question, the answer should include at least three parts: an example, specific benefits and specific harmful effects. If the answer incorporates as many benefits as harmful effects, this would be a good strategy. Since there are 9 marks, aim for at least four of each, plus the example. There is no need to try to compare or oppose benefits and harmful effects because they may not fall in the same category, but grouping answer elements could certainly add to the quality of structure for this answer.

Here the discussion should be based on a real example, not a hypothetical scenario. A complete answer would include the transfer details (source of the gene and organism to which it is transferred). The named example should contain the name of an organism, but also the circumstances (for example, salt tolerance in tomato plants). There may also be a mark to describe what is a genetic modification: when the DNA/genotype of an organism is artificially changed, therefore affecting its phenotype/characteristics.

List of benefits may include: growth of crop where it would not naturally occur; providing more food; less costly; reduce the required farming area or expand the production using the same area; less threat to other ecosystems (for example, rainforest); crop may require less chemicals, so it is better for the environment.

List of harmful effects may include: organism may be released in the environment; it may affect food chains or other organisms; organism may mix with others (that is, cross-pollination); long-term effects are unknown; usually are used by large farmers, so smaller farmers cannot compete; some consumers may develop allergies; they may contaminate the soil.

Be prepared

- A good understanding of basic techniques is required, but be prepared to answer questions about the ethics of these techniques, as shown above.

Explain the process of translation. *[9]*

[Taken from standard level paper 1, time zone 1, May 2009]

How do I approach the question?

- This question has occurred many times in past IBO exams, along with others about transcription and translation. You should fully differentiate and understand these processes, and be prepared to choose to write answers about them.
- The answer can begin with short statements to define translation and where it occurs (in the cytoplasm).
- The role and required composition details of each substance should be stated: mRNA, tRNA, ribosome.
- The distinction between codons, anticodons, amino acids and polypeptide should be clear.
- Building an answer on the chronological order of events is always a good strategy.

Which areas of the syllabus is this question taken from?

- The explanation of the process of translation (3.5.4)

This answer achieved 2/9

1 There is confusion between tRNA and anticodons ...
2 amino acids and ...
3 polypeptides.

There is a lot of confusion between the elements involved in translation. To write a valid answer, there should be a clear understanding of what mRNA, tRNA, codons, anticodons, amino acids and polypeptides are. A general idea of the process is present, but cannot be expressed clearly and logically. This answer does not contribute anything towards quality marks.

mRNA, that was produced in transcription, attaches to a ribosome ✓ in the cytoplasm ✓ of a cell. Free tRNA nucleotides **1** then attach to their complementary base pairs (A+U, C+G). Once attached with hydrogen bonds, the anticodon triplet coding for an amino acid can then join to the one next to it with a peptide linkage. Two nucleotides can be attached at a time. **2** Once the amino acid is formed, **3** it leaves the ribosome, eventually forming a chain of amino acids, known as a polypeptide chain.

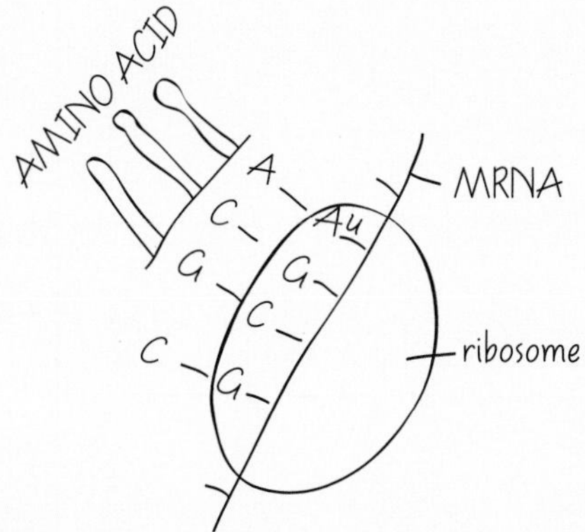

This answer achieved 5/9

1 The term "ribosome" should have been used here.
2 This happens before.

A general idea of the process is present, but the appropriate terminology is not always used and the student has difficulty relating the events in a sequential or chronological order; this affects the quality marks, explained in chapter 4.

Translation occurs in the cytoplasm. ✔ It uses the mRNA strand to create amino acid chains. ✔ The two protein ❶ subunits bind to the mRNA strand. Then the tRNA with the complementary anticodon attaches to the mRNA. ✔ After this, the amino acid which the codon represents attaches to the tRNA. ❷ A second tRNA "sticks" next to the first one ✔ and the protein ❶ moves down one codon. The amino acid from the first tRNA passes to the second and the first is released. The second tRNA now acts as the first one and does the same thing. Each time the protein ❶ will move down the mRNA ✔ and the amino acid chain becomes longer. This goes on until it reaches the stop codon which therefore releases both the protein subunits and the amino acid chain.

This answer achieved 9/9

1 This is not required for SL. No marks are allocated in markschemes for further knowledge than required in the syllabus, out of fairness to other students.

This answer relates the sequence of events correctly, using the appropriate terminology. It is presented clearly with a logical sequence, contributing to the quality marks.

Translation is the process by which amino acids are attached to create polypeptide chains. ✔ Translation occurs in the ribosomes, ✔ the site of protein synthesis. An mRNA enters with a code for a polypeptide chain ✔ transcribed from DNA. ✔ Three nucleotides in the mRNA sequence are referred to as a codon. ✔ When an enzyme reaches the "start codon" ❶ on the mRNA sequence it begins the process. tRNA molecules arrive at the ribosome carrying amino acids ✔ specific to the nucleotide sequence on the tRNA. A tRNA with an anti-codon that matches the next codon ✔ on the mRNA sequence (for example AGC matching with UCG) will attach itself to the mRNA, drop its amino acid, then leave. Each new amino acid that is brought on by a tRNA molecule is attached to the previous, ✔ creating a chain of amino acids known as a polypeptide ✔ or a simple protein. This process will continue until the stop codon on the mRNA is reached. At this point the tRNA will halt the bringing of amino acids and the newly made protein will leave the ribosome and fulfill its function.

Explain the consequences of altering a DNA base in the genome of an organism. *[8]*

[Taken from standard level paper 2, time zone 2, May 2009]

How do I approach the question?

An explanation requires an account of reasons and mechanisms. An ideal answer requires a brief outline of transcription and translation as well as an account of how altering a DNA base could impact both processes. You might take a stepwise approach. For any extended-response question, you are advised to include a specific example. The relevant example from the syllabus is that of sickle-cell anemia.

Which areas of the syllabus is this question taken from?

- Transcription (3.5.2)
- Translation (3.5.4)
- Gene mutation (4.1.3–4.1.4)

This answer achieved 3/8

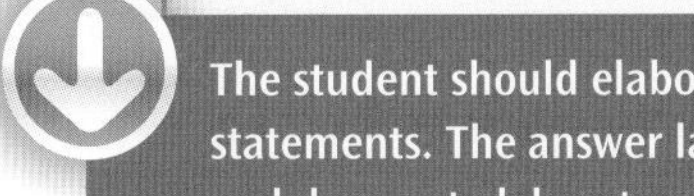

The student should elaborate on many of the statements. The answer lacks accurate details and does not elaborate on mechanisms.

The normal sequence of a codon for an Hb^A or normal blood may be CTG. This means that when it is translated, the sequence of the mRNA will be GAG. However, if there were to be a genetic mutation in the codon and CAC was being translated instead ✓ of CTC, the mRNA would have to use uracil to pair with the base adenine as there is no thymine in RNA. This change of base occurs due to a mutation ✓ and it results in a change in the phenotype of the cell. It is now Hb^S meaning it is a sickle cell. ✓ Because the genes were mutated, the resulting base was changed and the cell was affected.

This answer achieved 5/8

1 The student uses specific language to refer to the mRNA codon.

The student's answer is sparse and has focused on specifics alone. The student would gain marks by elaborating on the general mechanisms.

When a base is altered within a gene, this is known as gene mutation. In the sickle cell anemia ✓ gene, the sixth DNA codon is CTC. When transcription occurs, mRNA is created with the codon GAG and this leads to the amino acid glutamic acid being added to the growing polypeptide at the ribosome. A base substitution that changes ✓ the sixth mRNA codon to GUG ✓ 1 means that valine ✓ gets added instead. The result is sickle shaped blood cells. ✓

This answer achieved 8/8

An explanation requires an account of mechanisms and the student says very little about transcription and translation.

The student gains most of the marks through broad knowledge of the specific example.

The consequence of altering DNA base pairing in a genome can be shown using the example of sickle cell anemia. ✔ This occurs through a process called base mutation. This happens when base pairings are changed in which the triplet bases within the gene are altered at the 6th codon. ✔ The codon is changed from GAG to GTG. ✔ This change in the genome causes a change in the phenotype. This is due to the fact that a different amino acid is added: valine instead of glutamic acid. ✔ In this case, it causes a change from the normal Hb^A allele into the Hb^S allele ✔ which can lead to sickling of blood cells which shortens the life span of the cells. This results in anemia. ✔

Note that if a base sequence change occurs outside of a gene, there is likely to be no effect. ✔

Sometimes too, even if the change has occurred within a gene on the codon, no effect may occur because the code is degenerate. ✔ This means that two different codons can code for the addition of the same amino acid.

Meiosis

Required definitions for this section:

- **diploid nucleus**—a nucleus containing all homologous chromosomes in pairs, represented by **2N**
- **haploid nucleus**—a nucleus containing only one of each homologous chromosome, represented by **n**
- **homologous chromosomes**—chromosomes with the same length and structure that carry the same genes, each at the same locus.

You should know:

- the general process of meiosis
- the consequence of a non-disjunction
- the principle of karyotyping.

You should be able to:

- state that meiosis is a reduction division of a diploid nucleus to form haploid nuclei
- outline the process of meiosis
- explain that non-disjunction can lead to changes in chromosome number
- state that chromosomes are arranged in pairs according to their size and structure in karyotyping
- state that karyotyping is performed using cells collected by chorionic villus sampling or amniocentesis, for pre-natal diagnosis of chromosome abnormalities
- analyse a human karyotype to determine gender and whether non-disjunction has occurred.

Examples

1. Which phase of cell division is photographed in order to make a karyotype?
   A. Anaphase of mitosis
   B. Anaphase I of meiosis
   C. Metaphase of mitosis
   D. Metaphase II of meiosis

Amniocentesis or chorionic villus sampling harvests cells from tissues where cells multiply, therefore mitosis is involved. The best time to take the photograph is in metaphase (chromosomes condensed, easily seen), and therefore metaphase of mitosis (choice C) is the correct answer.

2. The following diagram represents a karyotype.

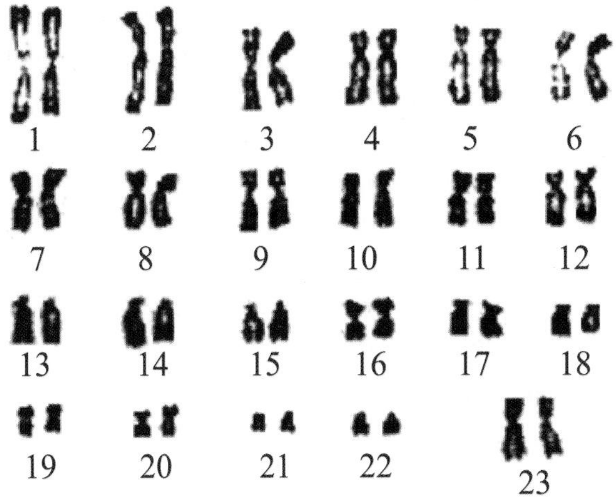

[*Source:* www.ds-health.com/trisomy.htm. Reproduced with permission.]

What can be concluded from the karyotype provided?
   A. There was non-disjunction during meiosis in the mother.
   B. There was non-disjunction during meiosis in the father.
   C. The fetus is male.
   D. The fetus is female.

This karyotype shows two long chromosomes for pair 23, that is, two X chromosomes, and the fetus is female. Choice D is the correct answer, and choice C is incorrect. To validate that choices A and B cannot present a correct answer, you can check if there are 46 chromosomes: this is the case, and a non-disjunction is ruled out.

Meiosis (continued)

Be prepared

- You should be able to distinguish clearly between the processes and the outcomes of mitosis and meiosis. The purpose of mitosis is to produce new identical cells, whereas the purpose of meiosis is to produce cells for reproduction (gametes).
- The names of the stages should be included in an outline of the process of meiosis. You must also mention the pairing of homologous chromosomes and crossing-over, followed by two divisions, which results in four haploid cells.
- Since meiosis involves different possibilities of distributing homologous chromosomes carrying genes and their alleles in reproductive cells, keeping track of combinations of alleles through meiosis and fertilization is the key to the understanding of theoretical genetics.
- You should be able to recognize the different stages of the two meiotic divisions on photographs or drawings.
- You must use Down syndrome (trisomy 21) to explain that non-disjunction can lead to changes in chromosome number, but examples with other chromosome abnormalities can be presented in exams for analysis.
- You must use your knowledge of gender determination described in the next section to analyse gender on a karyotype.
- Homologous chromosomes are likely to carry different alleles for some of their genes.

Theoretical genetics

Required definitions for this section:

- **carrier**—an individual that has one copy of a recessive allele that causes a genetic disease in individuals that are homozygous for this allele
- **codominant alleles**—pairs of alleles that both affect the phenotype when present in a heterozygote
- **dominant allele**—an allele that has the same effect on the phenotype whether the individual genotype is homozygous or heterozygous
- **genotype**—the alleles of an organism
- **heterozygous**—having two different alleles of a gene
- **homozygous**—having two identical alleles of a gene
- **locus** (plural: **loci**)—the particular position of a gene on homologous chromosomes
- **phenotype**—the characteristics of an organism
- **recessive allele**—an allele that only has an effect on the phenotype when present in the homozygous state
- **sex linkage**—the mode of transmission of genes carried on the sex chromosomes, mainly the X chromosome
- **test cross**—testing a suspected heterozygote by crossing it with a known homozygous recessive.

Useful terminology for this section:

- **monohybrid cross**—genetic cross between individuals where only one characteristic is considered (for example, fur colour)
- **multiple alleles**—type of inheritance for one characteristic due to more than two different alleles for the same gene, spread between members of the population.

You should know:

- the general terminology used in theoretical genetics
- the mechanism of monohybrid crosses
- inheritance involving multiple alleles
- inheritance involving sex linkage.

You should be able to:

- define genotype, phenotype, dominant, recessive and codominant alleles, locus, homozygous, heterozygous, carrier, test cross and sex linkage
- determine the genotypes and phenotypes of the offspring of a monohybrid cross using a Punnett grid
- state that some genes have more than two alleles (multiple alleles)
- describe ABO blood groups as an example of codominance and multiple alleles
- explain how the sex chromosomes control gender by referring to the inheritance of X and Y chromosomes in humans
- state that some genes are present on the X chromosome and absent from the shorter Y chromosome
- describe the inheritance of colour blindness and hemophilia as examples of sex linkage

Theoretical genetics (continued)

- explain that a human female can be homozygous or heterozygous with respect to sex-linked genes
- explain that female carriers are heterozygous for X-linked recessive alleles
- predict the genotypic and phenotypic ratios of offspring of monohybrid crosses involving any of the patterns of inheritance mentioned in this section
- deduce the genotypes and phenotypes of individuals in pedigree charts.

Examples

1. What evidence is given in the pedigree chart below to establish that the condition is caused by a dominant allele?

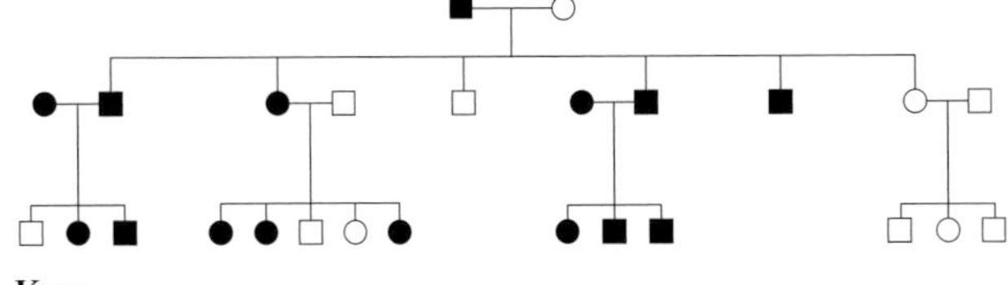

Key:
□ unaffected male　affected male ■
○ unaffected female　affected female ●

A. Two unaffected parents have unaffected children.
B. Two affected parents have affected children.
C. An affected parent and an unaffected parent have affected children.
D. Two affected parents have an unaffected child.

This question will be time consuming and difficult if you try to work out the pedigree, or if you let the key showing females and males distract you: gender is not in the answer choice. Instead, as in many genetics problems, start with the recessive allele. The recessive character can "skip generations", meaning that an individual with the recessive character might have parents who do not show the characteristic. The unaffected male (white square) from the third generation on the left shows this, and it demonstrates that the unaffected condition is recessive and therefore the affected condition is dominant. Choice D is therefore the correct answer.

2. If an organism that is homozygous recessive for a trait is crossed with a heterozygote, what is the chance of getting a homozygous recessive phenotype in the first generation?

A. 0%
B. 25%
C. 50%
D. 100%

The homozygous can only give recessive alleles for the trait. Half of the heterozygote's gametes would carry the dominant allele, the other half the recessive one, so half the offspring would be heterozygous and half homozygous recessive.

Homozygote's gametes: all a

Heterozygote's gametes: $\frac{1}{2}$A, $\frac{1}{2}$a

Crossing result (genotypes): $\frac{1}{2}$Aa, $\frac{1}{2}$aa

The correct answer is 50% (choice C).

Be prepared

- You must make sure that you can recognize the type of inheritance when you are given a genetics problem. Is it a monohybrid cross, multiple alleles, sex linked and so on?
- Appropriate notation for dominant and recessive alleles are upper-case and lower-case letters, respectively (for example, A and a).
- Appropriate notation for codominant alleles is a main letter relating to the gene and a suffix relating to the allele, both upper case, such as C^R and C^W for codominant red and white flowers, or Hb^A and Hb^S for normal red blood cells and sickle cells.
- Appropriate notation for ABO blood group alleles is I^A, I^B and i.
- Appropriate notation for dominant sex-linked alleles for colour blindness and hemophilia is X^B and X^H, whereas it is X^b and X^h for the recessive alleles.
- You should label parental genotypes, gametes, and both offspring "genotype" and "phenotype" when using a Punnett grid, unless stated otherwise.
- Some books use the term "Punnett square" instead of "Punnett grid", and you can use either in your answers.
- You should choose letters representing alleles with care to avoid confusion between upper and lower case when they are not provided. Although seen in textbooks, using different letters to designate the alleles of a gene is not recommended, as it may lead to confusion.
- Genotypes taken from publications used in data-based questions may involve a combination of many letters and/or numbers to designate one single gene, such as the breast-cancer-related gene BRCA1.
- Sometimes two alleles are distinguished by the symbol $^+$ for the "wild type" (the usual phenotype present in the population).

A genetic cross was made between pure-breeding snapdragon plants with red flowers and pure-breeding snapdragon plants with white flowers. The cross produced F_1 offspring that had only pink flowers. When the F_1 plants were self-pollinated, the resulting F_2 generation had some red, some white and some pink flowers.

(a) (i) Identify the relationship between the red and white alleles for flower colour. *[1]*

(ii) Deduce the genotype of the F_1 plants. *[1]*

(iii) Construct a Punnett grid to show the cross between two F_1 plants. *[2]*

(iv) Deduce the proportion of the different phenotypes of the F_2 offspring. *[1]*

[Taken from standard level paper 2, November 2006]

How do I approach the question?

- You can invent your own symbols for the alleles, but they should be consistent with the codominant nature of the alleles.
- When drawing the Punnett grid, in part (iii), you should clearly show gametes.
- In part (iv), when showing phenotypes, either ratios or percentages can be shown.

Which areas of the syllabus is this question taken from?

- Terminology used in theoretical genetics (4.3.1)
- Monohybrid cross and Punnett grid (4.3.2)
- Codominant alleles (4.3.4)

This answer achieved 3/5

1 The symbols chosen indicate that the student does not fully understand the conventions for naming alleles.

(i) They are codominant alleles ✓

(ii) heterozygous ✓

(iii)

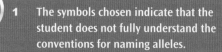

	R	i **1**
W	RW	Wi
I	Ri	ii

(iv) 2:1:1. In F_2, there would be two red flowers one white flower and one pink flower ✓

This answer achieved 5/5

1 The student has used conventional symbols for codominant alleles and has drawn the Punnett grid correctly.

(i) They are codominant alleles. ✔

(ii) If C^R represents the red allele and C^W represents the white allele, then the genotype of the F_1 plants would be C^RC^W. ✔

(iii)

	C^R	C^W ✔ 1
C^R	C^RC^R	C^RC^W ✔
C^W	C^RC^W	C^WC^W

(iv) 1 red : 2 pink : 1 white ✔

Outline the role of sex chromosomes in the control of gender and inheritance of hemophilia. *[6]*

[Taken from standard level paper 2, time zone 1, May 2009]

How do I approach the question?

"Outline" requires that you provide a brief summary. This would involve summarizing the chromosome combination in males and females, summarizing the concept of sex linkage, including why X-linked recessive conditions are more common in males, and finally outlining the inheritance of hemophilia as an example of an X-linked recessive condition. Supplementing answers with Punnett grids is a recommended strategy.

Which areas of the syllabus is this question taken from?

- Sex chromosomes and sex linkage (4.3.5–4.3.8)

This answer achieved 3/6

1 The student needs to elaborate on the points made, as key ideas are left unstated.

Sex chromosomes have a role in controlling gender. The Y chromosome has a gene on it that causes males to develop. XY makes a male ✔ and XX makes a female. ✔ Hemophilia is a sex-linked trait ✔ because it is on the X chromosome. An offspring has two chromosomes: one from the mother and one from the father. If the mother is a carrier of hemophilia, then the son will be affected by it. 1

This answer achieved 6/6

1 The student is accompanying the answer with a Punnett grid. It is recommended that, where relevant, students should support extended responses with diagrams.

The student has fully elaborated on all key points.

Gender is determined by the sex chromosomes. A mother produces gametes that only have the X chromosome. A father produces two kinds of gametes, those that have an X chromosome and those that have a Y chromosome. ✓ The Punnett grid below shows the possible offspring.

1	X	Y
X	XX	XY
X	XX	XY

If a person has two X chromosomes they will be female ✓ and if they have an X and Y chromosome they will be male. ✓ The grid above allows us to predict that half of the children will be males and the other half will be females.

Hemophilia is a sex-linked disease ✓ as its locus is on the X chromosome. Some sex-linked characteristics can be found on the Y chromosome. Hemophilia is also a recessive condition. If a woman possesses the hemophilia allele, she is most likely to have only one copy so she will be heterozygous and unaffected. ✓ In this case, we would call her a carrier. If a male has a copy of the hemophilia allele, he will be affected because there is only one X chromosome in males.

If a carrier mother has children with an unaffected father, the possible children are shown in the grid below. X^H will be used to represent the unaffected allele and X^h will be used to represent the affected allele.

	X^H	Y ✓
X^H	$X^H X^H$	X^H Y
X^h	$X^H X^h$	X^h Y

The result will be that half the daughters will be carriers and half will not have the allele. Half the sons will be affected ✓ and half will be unaffected.

10. Classification, ecology and evolution

Communities, ecosystems and biomes

Required definitions for this section:

- **community**—a group of populations living and interacting with each other in an area
- **ecology**—the study of relationships between living organisms and between organisms and their environment
- **ecosystem**—a community and its abiotic environment
- **habitat**—the environment in which a species normally lives or the location of a living organism
- **population**—a group of organisms of the same species that live in the same area at the same time
- **species**—a group or organisms that can interbreed and produce fertile offspring
- **trophic level**—the position of an organism in a food chain or a food web (for example, secondary consumer).

Useful terminology for this section:

- **abiotic**—the non-living components of an ecosystem
- **autotroph**—an organism that synthesizes its organic molecules from simple inorganic substances
- **consumer**—an organism that ingests other organic matter that is living or recently killed
- **detritivore**—an organism that ingests non-living organic matter
- **heterotroph**—an organism that obtains organic molecules from other organisms
- **saprotroph**—an organism that lives on or in non-living organic matter, secreting digestive enzymes into it and absorbing the products of digestion.

You should know:

- common definitions used in ecology
- trophic levels of organisms in food chains and food webs
- the principles of energy flow in a food chain and a food web
- the principle of nutrient recycling.

You should be able to:

- define species, habitat, population, community, ecosystem, ecology and trophic level
- distinguish between autotrophs and heterotrophs
- distinguish between consumers, detritivores and saprotrophs
- give three examples of food chains, each composed of at least four organisms
- deduce the trophic level of organisms in a food chain and a food web
- construct a food web containing up to 10 organisms, using appropriate information
- state that light is the initial source of energy for almost all communities
- explain the energy flow in a food chain
- state that energy transformations are never 100% efficient
- explain the reasons for the shape of pyramids of energy
- explain that energy enters and leaves ecosystems, but nutrients must be recycled
- state that saprotrophic bacteria and fungi (decomposers) recycle nutrients.

Examples

1. Slime moulds (*Acrasiomycota*) are protoctists. They feed on decaying organic matter, bacteria and protozoa.

   Which of these terms describes their nutrition?

   I. Detritivore
   II. Autotroph
   III. Heterotroph

   A. I only
   B. I and II only
   C. I and III only
   D. I, II and III

Communities, ecosystems and biomes (continued)

In this question, the names of organisms should not be a concern, since only their mode of nutrition is important for the answer and is described in the second sentence. (Further details about the classification of organisms are only required from students preparing for option F.) Detritivores are organisms feeding on non-living organic matter, so the term can apply to slime moulds since the question says that they feed on decaying organic matter. Option I is therefore true. Bacteria and protozoa are living organisms, and organisms that obtain organic matter from other organisms are heterotrophs. Therefore, option III is also true. The question excludes the mention that they synthesize their organic molecules from simple organic substances. Therefore slime moulds are not autotrophs, as option II suggests. The choice C, options I and III only, is therefore the right answer.

2. Why do food chains in an ecosystem rarely contain more than five organisms?
   A. Nutrients are recycled by the decomposers back to the producers.
   B. Nutrients are lost from the ecosystem when organisms die.
   C. The conversion of food into growth by an organism is not very efficient.
   D. Energy is recycled by the decomposers back to the producers.

Only choice C relates to different trophic levels. Choice A supports longer food chains, and choice B is not generally true as most ecosystems are closed. Choice D is not a correct statement since energy is not recycled.

Be prepared

- In the definition of ecology, the wording may look strange, but two elements/ideas are implied: (1) relationship between different organisms and (2) relationships between the organisms and the (abiotic) environment.
- Do not use the term "species" and "individual" interchangeably.
- Many examples of food chains are available in textbooks and on the Internet, but you have to make sure that they represent documented relationships between organisms.
- Organisms named in examples of food chains should be at the species or genus level, but you can also use common names instead of binomial names (for example, raccoon instead of *Procyon lotor*).
- Plankton is sometimes mentioned in food chains or food webs. It designates a collection of microscopic aquatic (fresh-water) or marine (salt-water) organisms, at the base of food chains or food webs, regardless of their position in the classification of organisms. Plankton is sometimes subdivided into phytoplankton (grouping autotrophic organisms) and zooplankton (grouping heterotrophic organisms).
- Although only named organisms are accepted in examples of food chains or food webs, you can use phytoplankton as producers and zooplankton as first levels of consumers. This is the only accepted exception for an aquatic or marine food chain also involving larger organisms.
- Use species that occur in the same ecosystem/biome (for example, caribou and camels are not in the same food web).
- Arrows should indicate the direction of energy flow in food chains or food webs (A $\rightarrow$ B indicates that A is being "eaten" by B). Arrows represent flows.
- Energy losses between trophic levels include material not consumed or material not assimilated, and heat loss through cell respiration.
- Units for pyramids of energy are $\text{kJ m}^{-2}\text{yr}^{-1}$.

The greenhouse effect

You should know:

- processes involved in the carbon cycle
- the relationship between atmospheric gases and the enhanced greenhouse effect
- the statement and implications of the precautionary principle
- consequences of global warming on arctic ecosystems.

You should be able to:

- draw and label a diagram of the carbon cycle to show the interaction of living organisms and the biosphere through photosynthesis, respiration, fossilization and combustion
- analyse the changes in concentration of atmospheric carbon dioxide using historical records

The greenhouse effect (continued)

- explain the relationship between rises in concentration of carbon dioxide, methane and oxides of nitrogen and the enhanced greenhouse effect
- outline the precautionary principle
- evaluate the precautionary principle as a justification for strong action in response to the threats posed by the enhanced greenhouse effect
- outline the consequences of a global temperature rise on arctic ecosystems.

Examples

1. The diagram below shows some of the links in the carbon cycle.

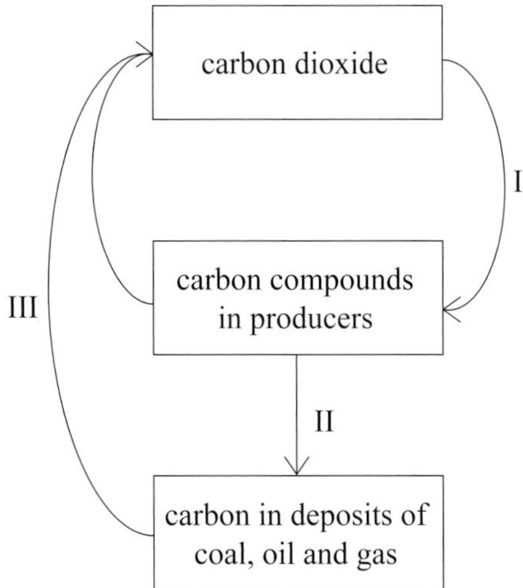

What processes are taking place at I, II and III?

	I	II	III
A.	photosynthesis	fossilization	combustion
B.	cell respiration	fossilization	greenhouse effect
C.	photosynthesis	decomposition	combustion
D.	cell respiration	decomposition	greenhouse effect

This question is based on the knowledge of the names of the processes taking place in the carbon cycle. Only photosynthesis enables the incorporation of CO_2 into organic compounds for process I. Migration of carbon from organisms into deposits is called fossilization. Choice A is therefore the right answer.

2. Explain how the emission of gases, both naturally and through human activity, can alter the surface temperature of the Earth. *[8]*

You must address the command term "explain", therefore giving an account of reasons and mechanisms. Your first step is to determine the nature of the problem that needs to be explained and to eliminate irrelevant areas. The question is therefore about relating greenhouse gases to the Earth's temperature and excludes other atmospheric problems such as ozone depletion and acid rain. To cover all the question's elements, you should name the gases and state which are emitted naturally and through human activity. You should also explain how this happens. Gases to be named, with their sources, are: carbon dioxide—increased by the combustion of fossil fuels; methane—produced by decomposition; nitrous oxides—produced by some industrial processes and released by volcanic activity. Short-wave radiation from the Sun enters the atmosphere, warms the Earth's surface and is released as long-wave radiation or heat. Gases trap this long-wave radiation, increasing the atmosphere's temperature: this is the greenhouse effect. The fact that the Earth's temperature has risen over the last century is global warming. You can mention that the greenhouse effect is normal, but that human activity has increased the amount of greenhouse gases, thus causing global warming.

Be prepared

- Do not confuse the effects of greenhouse gases on the temperature of the atmosphere with other atmospheric problems, including ozone depletion.
- Carbon dioxide, methane and oxides of nitrogen are greenhouse gases, but some other gases released in the atmosphere, such as chlorofluorocarbons (CFCs), can contribute to both the enhanced greenhouse effect and ozone depletion.
- When you draw the carbon cycle, it should show that photosynthesis uses CO_2 from the atmosphere, whereas respiration releases it, that carbon is transferred through trophic levels and sediments as organic matter, and that CO_2 is released to the atmosphere from the respiration of living organisms, including decomposers, and combustion of organic matter.
- Be aware that the greenhouse effect is a natural phenomenon, but that an **enhanced** greenhouse effect has been noticed from the second half of the 20th century.

The greenhouse effect (continued)

- In your explanations of the enhanced greenhouse effect, mention that incoming short-wave radiation in the atmosphere is re-radiated as longer-wave radiation, including infrared (IR) in the form of heat.
- Historical records to analyse atmospheric carbon dioxide levels are usually very well quoted examples in textbooks, such as the Keeling data from Mauna Loa, Hawaii, also available directly from the Internet.
- You are expected to be able to evaluate not only the general trend of CO_2 levels from data, but also the yearly fluctuation pattern.
- You should express clearly what the precautionary principle states: that, if the effects of a human-induced change would be very large, perhaps catastrophic, those responsible for the change must prove that it will not harm before proceeding.
- Your evaluation of the precautionary principle as a justification for strong action must assess implications and limitations.
- Outline how a global temperature rise might affect arctic ecosystems by citing the following: increased rates of decomposition of detritus previously trapped in permafrost; expansion of the range of habitats available to temperate species; loss of ice habitat; changes in distribution of prey species affecting higher trophic levels; and increased success of pest species, including pathogens.

Populations

You should know:

- the effect of natality (birth rate), immigration, mortality (death rate) and emigration on population size
- the interpretation of a population growth curve
- factors that set limits to population increase.

You should be able to:

- outline how population size can be increased by natality and/or immigration and decreased by mortality and/or emigration
- draw and label a graph showing a sigmoid (S-shaped) population growth curve
- explain the reasons for the three phases of a population growth curve.

Examples

1. The following diagram shows part of a food web from Yellowstone Park.

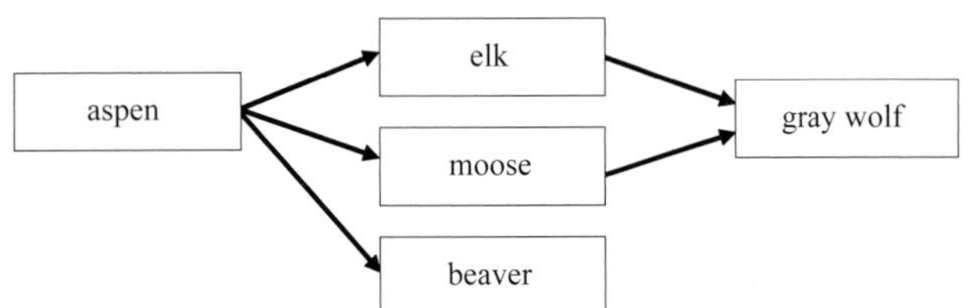

What would be the short-term effects on the populations of the other species if the gray wolf were exterminated?

	Beaver	Moose	Elk	Aspen
A.	increase	decrease	increase	increase
B.	decrease	decrease	decrease	decrease
C.	increase	increase	decrease	increase
D.	decrease	increase	increase	decrease

Predation is a factor limiting the increase of a population. A decrease in gray wolf predation would increase its preys' populations, therefore increasing elk and moose populations. This would in turn increase the competition, another population limiting factor, on the beaver population, thus reducing the beaver population. It is likely that there would be more primary consumers feeding on aspen, therefore decreasing its population. The only choice meeting these conditions is D.

2. The diagram below shows a population growth curve.

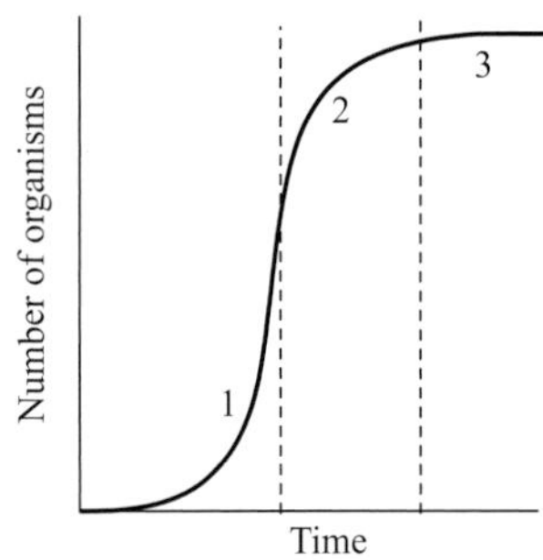

At which time in the population growth curve does the population size begin to decline?

Populations (continued)

A. Between the times marked 1 and 2

B. During the time marked 2

C. Between the times marked 2 and 3

D. The graph does not show a time when population size declines.

This question assumes knowledge of factors affecting the size of a population and the understanding of the difference between an absolute variation in numbers and the variation in **rate** (see "Be prepared" below). The curve in phases 1 and 2 always shows a growth, but the **rate of growth** increases in 1, whereas it slows down in 2 and 3 to become nil at the end of 3. Although the growth stops, the population size never declines, so choice D is the right answer.

Be prepared

- Natality, immigration, mortality and emigration are often expressed as rates in number of individuals per unit of time.
- Three factors that set limits to population increase are the resources (be able to name food, water and territory), predation and disease.
- The three phases of a sigmoid (S-shaped) population growth curve are the exponential growth phase, the transitional phase and the plateau phase, respectively.
- A curve with a positive slope on a graph (going up towards the right [/]) always means an increase, whereas a negative slope (going down towards the right [\]) means a decrease; a line parallel to the *x*-axis represents a stable condition.
- Since population growth is often expressed as a rate rather than as absolute numbers, use the appropriate terminology, especially when answering data-based questions. A positive slope becoming less steep means that the actual population is still increasing, but that the **rate of increase** (angle of slope) is decreasing!
- The **rate of increase** of a population decreases along the phases of a population growth curve because the factors limiting population growth become more important.

Evolution

Required definition for this section:

- **evolution**—cumulative change in the heritable characteristics of a population causing species change.

Useful terminology for this section:

- **natural selection—**the survival of better-adapted individuals over less well-adapted individuals in the same population in the same environment.

You should know:

- the evidence for evolution
- the importance of sexual reproduction for natural selection
- how natural selection leads to evolution.

You should be able to:

- outline the evidence for evolution, based on fossils, selective breeding and homologous structures
- explain the problem of too many offspring in a limited environment
- explain how sexual reproduction promotes variation in a species
- explain how natural selection leads to evolution
- explain two examples of evolution in response to environmental change, including antibiotic resistance in bacteria.

Examples

1. Which process tends to reduce variety within a population?

   A. Natural selection

   B. Random fertilization

   C. Independent assortment

   D. Crossing over

   Answer A is correct because natural selection reduces variety as only the best-adapted individuals in a population survive to reproduce. The reproductive events named in B, C and D all lead to greater variety of offspring in a population.

2. Explain how sexual reproduction can lead to variation in a species. *[3]*

   First of all, it should be mentioned that sexual reproduction allows characteristics from both parents to appear in

Evolution (continued)

offspring, so that offspring will always be different from their parents. During gamete production, variation in the genetic combination of homologous chromosomes in both parents will always be ensured by two events: (1) crossing over between homologous chromosomes during prophase I of meiosis; and (2) independent assortment of paired genes on homologous chromosomes during metaphase I of meiosis. Finally, the random chance of which sperm fertilizes which ovum also contributes to variation in a species.

Be prepared

- Evolution is the meeting ground for concepts from different core topics of the biology syllabus. These include "The chemistry of life" (DNA information), "Genetics" (chromosomes, genes, alleles, mutations and meiosis) and, of course, "Ecology and evolution". As a result, there are many chances to relate different ideas that may have been learned weeks and months apart.
- You should be able to recall the concept of a gene and to distinguish between chromosomes, genes and alleles.
- You should be able to recall what a mutation is and how it can occur.
- You should remember how mitosis varies from meiosis.
- In your studies of evolution, or any other topic in biology, try to think beyond what you are learning at the moment and be alert to connections with other knowledge previously learned. Not only will you be reinforcing through review, but you will also be enriching your overall understanding and appreciation of biology.

Classification

You should know:

- the principles of the binomial system of nomenclature
- the levels in the hierarchy of taxa
- distinctive external features of phyla of plants and animals
- the use of a dichotomous key.

You should be able to:

- outline that the binomial system of nomenclature consists of naming each species with a unique two-word combination in Latin, the first referring to the genus, the second referring to the species
- list seven levels in the hierarchy of taxa, using an example from two different kingdoms for each level
- distinguish between the following different phyla of plants, using simple external recognition features: Bryophyta, Filicinophyta, Coniferophyta and Angiospermophyta
- distinguish between the following different phyla of animals, using simple external recognition features: Porifera, Cnidaria, Platyhelminthes, Annelida, Mollusca and Arthropoda
- apply and design a dichotomous key for a group of up to eight organisms.

Examples

1. *Pseudolarix amabilis* produces seeds but not flowers. *Physcomitrella patens* has leaves but not roots. To which groups do they belong?

	Pseudolarix amabilis	*Physcomitrella patens*
A.	Coniferophyta	Filicinophyta
B.	Filicinophyta	Angiospermophyta
C.	Coniferophyta	Bryophyta
D.	Angiospermophyta	Coniferophyta

Choice B is excluded because filicinophytes do not produce seeds. Although coniferophytes and angiospermophytes produce seeds, only angiospermophytes produce flowers, therefore eliminating choice D. Choices A and C are valid for *Pseudolarix amabilis*, but filicinophytes possess true leaves and roots. Choice C is therefore the correct answer.

Classification (continued)

2. Which of the organisms A–D, identified by the key below, represents an annelid?

(1)	Shows bilateral symmetry	go to 2
	Does not show bilateral symmetry	Cnidaria
(2)	Has a segmented body	go to 3
	Does not have a segmented body	go to 4
(3)	Has jointed legs	A
	Does not have jointed legs	B
(4)	Has a shell	C
	Does not have a shell	D

This question was based on the external recognition of named features of organisms belonging to the phylum Annelida, but also on the ability to apply a dichotomous key. Annelida show bilateral symmetry, leading to a segmented body (2), but for (3) do not have jointed or articulated legs (these would be Arthropoda). Therefore B is the right answer.

Be prepared

- Species with the same last binomial name are usually not closely related since they belong to at least different genera (genuses), as in *Castor canadensis* (beaver) and *Solidago canadensis* (Canada goldenrod).
- Classification occupies only a small amount of the biology syllabus. Do not be disturbed by the presentation of organisms belonging to groups that you were not supposed to know in exam papers: the question will often be testing something else than this area of the classification!
- Although not specified in the subject guide, be aware that the kingdoms are: Prokaryotes, Protists, Fungi, Plants and Animals (students preparing for option F should look at chapter 18 for further details).

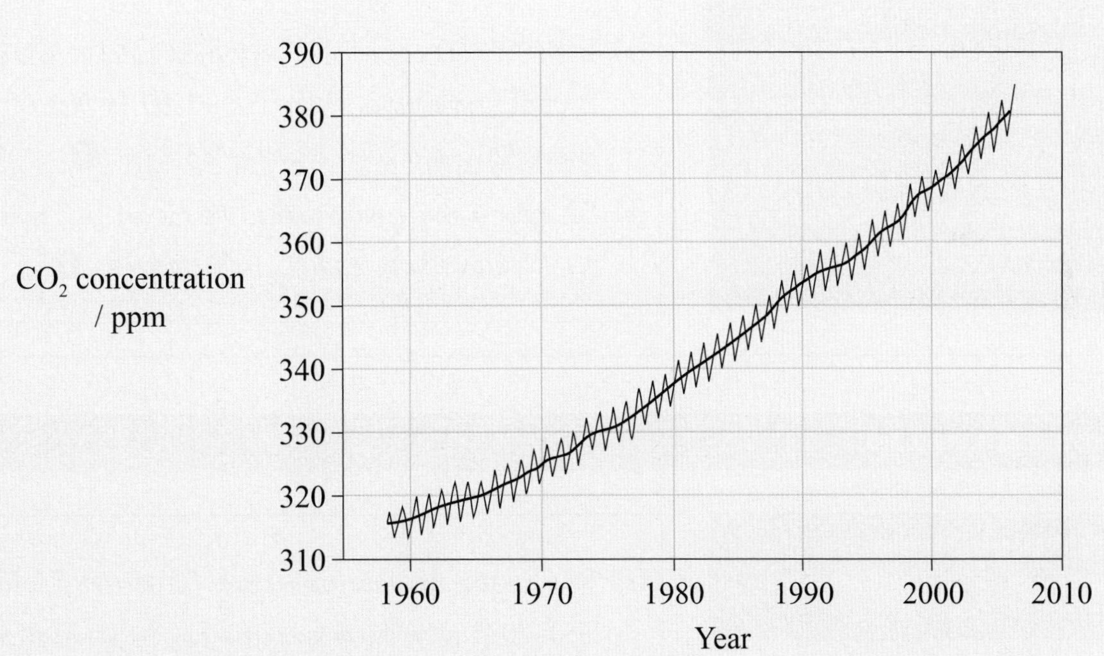

[*Source*: Adapted from Dr Tans, P. Full Mauna Loa CO_2 record. Trends in atmospheric carbon dioxide. NOAA Earth Systems Research Laboratory. Copyright © Dr P Tans. Reproduced with permission.]

(a) Explain the observed changes in atmospheric CO_2 concentrations from 1960 to 2005. *[3]*

(b) Outline the precautionary principle. *[2]*

[Taken from standard level paper 2, time zone 1, May 2009]

How do I approach the question?

(a) This part requires an explanation. Although describing the changes could probably earn a mark, you should focus on the causes, mainly the release of CO_2 by human activities, and give an example, such as combustion of fossil fuels and deforestation. You can also mention that there is a yearly variation due to the increase of photosynthesis in the northern hemisphere.

(b) This question is another example of how you can increase your total score by including additional knowledge. The two elements of the precautionary principle are: (1) human-induced change can be catastrophic and (2) those responsible for the change must prove that it will not harm before proceeding. This would usually be enough to gain the two marks, but it would be a good strategy to add an example. Also, mentioning that this is the reverse of the usual practice (where those who are concerned about the change must have proven that it would harm) adds a finishing touch to the answer.

Which areas of the syllabus is this question taken from?

- Analysis of atmospheric carbon dioxide levels using historical records (5.2.2)
- The outline of the precautionary principle (5.2.4)

This answer achieved 1/5

1 This answer is limited to the description of the trend and does not provide any explanation.

2 This answer is irrelevant, perhaps because the precautionary principle was not covered in class. Make sure that you can retrieve all elements mentioned in the "You should be able to" from this book in your class notes.

(a) In 1960, the CO_2 concentration was low and started increasing. ✓ 2005 CO_2 concentration is higher than in 1960. In 1960, CO_2 reached 320 ppm, in 2005, it went up to 370 ppm. **1**

(b) CO_2 may rise more, greenhouse effect will rise and it will be hard to stop it. There will be too much CO_2 in the atmosphere. **2**

This answer achieved 3/5

1 Besides the trend description, annual fluctuations add to the information provided in this answer.

1 The answer is limited to a description and does not provide an explanation, as stated in the question.

2 The general idea is stated, but the catastrophic dimension is not mentioned, nor is the fact that it is those responsible who must demonstrate that it will cause no harm.

(a) There is a general increase in atmospheric CO_2 concentration from 1960–2005. ✓ Although there are annual fluctuations ✓ **1** of atmospheric CO_2 concentration, the overall rise is from 315 ppm in 1960 to 381 ppm in 2005. **1**

(b) The precautionary principle states that preventative measures should be taken in order to reduce greenhouse gas emissions, despite a lack of sufficient evidence. ✓ **2**

This answer achieved 5/5

1. The extent of the increase is stated; this is better than only quoting values.
2. Increased production of CO_2 is linked with causes: this addresses the command term "explain".
3. This demonstrates a thorough understanding of the precautionary principle that goes beyond the recall of factual information.

For both sub-questions, this student has expanded the answers to demonstrate an understanding of the issues, rather than a mere recall of information.

(a) Atmospheric CO_2 concentration increased by about 61 ppm (1) from 1960–2005. ✔ ... This staggering change can be largely explained by technological advances ✔ and changes: technology based on the combustion of fossil fuels ✔ (which produces CO_2) (2) has grown and become cheaper (production costs have lowered with advanced technology) in this time, so CO_2 conc. has increased. Cars are more common & cheaper, for example, and factories have become more abundant as industries grow.

(b) The precautionary principle is essentially a theory which encourages preventive action when prospective danger is perceived. It is seen to be safer to take action to **prevent** global warming, ✔ for example, rather than waiting for a danger to be **proven** as harmful (by this time, damage could be done/irreversible). (3) In short, those who go by the precautionary principle will take action to prevent perceived dangers in the future unless this action is **proven** ✔ to be inconsequential or unhelpful.

(a) Define the terms species, population and community. *[3]*

(b) Explain the shape of the pyramids of energy that are constructed by ecologists to represent energy flow in an ecosystem. *[3]*

[Taken from standard level paper 2, time zone 2, May 2009]

How do I approach the question?

(a) This part is a straightforward question based on knowledge of terminology. Basic knowledge of facts and terminology is often overlooked by many students when preparing for the exam papers, although it is an easy way to increase your total score. This is why you can find definitions at the beginning of each section of this book, and those provided in this chapter are valid answers for this question. You can assume here that the definition of each term is worth 1 mark, since the total is out of 3 marks. To gain the marks, it is important that you indicate details and circumstances (for example, "living and interacting", "in the same area", and so on). You should be accurate with the terms that you use: species does not mean individuals or populations, as in some of the following examples.

(b) Although you can use a labelled diagram to answer this question, the command term "explain" requires an indication of the energy flow and the processes linking each trophic level. You should include the word "trophic level" in your answer and show the different trophic levels. It is also important that you mention that energy passes from one level to the other. The reasons for energy "losses" between each level are also required: material not consumed, material not assimilated and energy loss through cell respiration. An indication of the importance of the losses (80–90% lost; 10–20% passed on) can also be an answer element.

Which areas of the syllabus is this question taken from?

- Definitions of ecology terms (5.1.1)
- Explanation for the reason of the shape of pyramids of energy (5.1.12)

This answer achieved 2/6

1. The word "species" cannot be used instead of "individuals of the same species".
2. It should be "populations".
3. There is no statement that energy is transferred from one trophic level to the next, and the trophic levels are not identified.
4. There is no explanation as why there is less energy to pass on.
5. The student does not state clearly that, since some material consumed is not digested, it cannot be passed on to the next trophic level.

(a) Species: A group of organisms which can interbreed and produce offspring which is fertile. ✓

Population: is the recorded number of species ❶ at a given time in a given location.

Community: is a group of different species ❷ interacting with each other and living together.

(b) There is less energy because as you move up trophic levels, there is less energy to be transferred between the levels. ❸ Producers produce a substantial amount of energy but when energy flows to a different trophic level, there is less energy to pass on and thus the energy decreases. ❹ This energy is lost as heat ✓ and other waste such as undigested plant matter. ❺

This answer achieved 4/6

1. The idea of energy transfer is written, although "between trophic levels" is missing.
2. The magnitude of transfer is stated.

1. "Individuals of the same species" needs to be mentioned here.
2. A diagram showing the flow of energy through labelled trophic levels would have been useful in this answer.

(a) Species: group of organisms that interbreed or have the potential to interbreed with each other to produce fertile offspring. ✔

Population: a group of same species ❶ living and interacting with each other in the same area at the same time.

Community: a group of populations living and interacting with each other in an area. ✔

(b) At each level of the pyramid, there will be less energy than at the level before. This is because not all energy is transferred. Much of it leaves before being able to be transferred. ❶ Energy leaves a level in the form of heat, ✔ waste, etc. The efficiency of energy transfer is lower than 20% between trophic levels. ✔ ❷ ❷

This answer achieved 6/6

1. Concepts such as production, respiration and biomass are mentioned.
2. The explanation relates cause to effect.

✓ The appropriate terminology used in these definitions makes the distinction between a correct and an incorrect answer. The explanation is clear and relates reasons and consequences.

(a) **Species:** Group of organisms with similar characteristics capable of interbreeding with each other to produce fertile offspring. ✔

Population: A group of organisms, of the same species, living and interacting with each other at the same place and at the same time. ✔

Community: A group of different populations living and interacting with each other and their environment. ✔

(b) Producers are at the base of the food chain. They fix energy from the sun directly and are therefore the most numerous and would therefore have the highest relative biomass. Some of the energy fixed by plants is used by the plant for respiration ❶ so net production is less than gross production. Only net production is available to the next trophic level which is why the bars on the pyramid get smaller. ❷ Only between 10 to 20% of gross production ends up being in the next trophic level. ✔ Besides heat losses from respiration, ✔ some of the energy is lost as undigested material in the form of feces or fur. ✔ The energy in these forms of matter is not passed on to the next trophic level.

Outline how antibiotic resistance in bacteria arise in response to environmental change. *[5]*

[Taken from higher level paper 2, time zone 1, May 2009]

How do I approach the question?

Your answer has to address both the antibiotic resistance and the environmental change. You have to relate your answer to biological mechanisms such as gene transmission and natural selection. You can use an example, but this should be a limited part of your answer and support your explanation.

Antibiotic resistance for a specific antibiotic is due to the presence of a gene or genes. Depending on the gene/combination of genes, resistance can exhibit varying degrees. It can pass from one generation of bacteria to the other if the gene is transmitted. The way to transmit it is through binary fission, but bacteria can also exchange plasmids and genetic material through conjugation (plasmids are covered along with genetic engineering in the syllabus, showing the interrelation of topics). A mutation can also confer a specific antibiotic resistance by altering a gene, which can then be transmitted in the same manner.

The exposure of bacteria to antibiotics is the environmental change, and it is the factor for natural selection. When bacteria are exposed to the antibiotic, only the resistant ones (for the applied dosage) survive. Combined with the fact that bacteria reproduce very quickly and have a high rate of mutation, the entire population can become resistant to the antibiotic within a few generations, especially if the dosage of the antibiotic is not administered properly. You can quote that not finishing the quantity of prescribed antibiotics creates an environment that favours the reproduction of bacteria with resistance to the antibiotic. The same process can occur with different antibiotics, creating a multi-resistant bacteria strain, as happens with MRSA (Methicillin-resistant *Staphylococcus aureus*) in hospitals.

Which areas of the syllabus is this question taken from?

- Explain examples of evolution in response to environmental change (5.4.8)

This answer achieved 1/5

1 The answer is limited to the description of an example and does not provide an explanation, as stated in the question. It does not relate to any biological structures or processes, such as genes, gene transfer and so on.

Antibiotic resistance in bacteria can arise in response to environmental change. A way in which this is achieved is through using antibiotics improperly. When a patient fails to take the recommended dose for a period of time, ✓ the bacteria left over will begin to reproduce exponentially with resistance to the prescribed antibiotic. (1) A patient who does this rises the probability for his/her community to contract this resistant bacteria. With increasing amount of patients doing this, increased amounts of antibiotics become non-useful for the treatment of infections. This is why it is important to improve use of antibiotics to reduce the population of resistant bacteria in the environment.

This answer achieved 3/5

1 The action of natural selection is well understood.

1 "Means" has to be specified (for example, plasmid transfer, conjugation and so on) to score marks.

2 The words "gene" or "allele" are required here.

Bacteria can develop resistance to antibiotics, or even multiple-resistant bacteria. A mutation, or entering of the gene that causes resistance into the bacteria by some means, (1) will occur in a bacteria. This bacteria will pass on the antibiotic resistance, (2) and so there will now be some bacteria with some antibiotic resistance and some without.

When a doctor or vet uses an antibiotic, it will kill all the bacteria that are not resistant to it. ✓ The result is that the remaining bacteria will be resistant, and reproduce, so now the entire population is resistant to that antibiotic. ✓ When doctors or vets switch to using a different or stronger antibiotic, it will again kill those not resistant to it, and so the remaining bacteria are now resistant to two kinds of antibiotics. The more antibiotics are used, the more resistant bacteria become to them, as the resistance is passed on to following generations ✓ until those resistant take over the population, as they can survive. (1)

This answer achieved 5/5

This answer articulates very well the biological processes of resistance, with the mechanisms of natural selection within a population, showing a thorough command of concepts, principles and terminology.

Antibiotics are chemicals that are produced by fungi such as penicillin and that prevent a bacterium from carrying out its metabolic processes and thus kill the bacterium. When a population of bacteria is introduced to an environment with high concentration of antibiotics, ✓ most of the organisms would die. After a while, however, the DNA in some bacteria might mutate and randomly lead to the development of an antibiotic resistance. Since the non-resistant bacteria would all be killed by the antibiotic ✓ nature of the environment, only the resistant ones could reproduce and inherit their resistance genes. ✓ Bacteria can also exchange genetic material via conjugation ✓ of their pilli and thus increase the spread with which the antibiotic resistant genes are spread throughout the population. Over time, the bacteria population would have evolved ✓ to be completely resistant against antibiotics of a certain type. When the environment was to change back to the initial, antibiotic-free one, the number of resistant bacteria would decrease slightly again.

11. Human physiology, part 1 (Digestion, transport and gas exchange)

Digestion

Required definitions for this section:

- **absorption**—passage of digestive products in molecular form through the lining of the digestive tract for eventual entry into the transport system
- **assimilation**—construction of macromolecules in body cells from absorbed amino acids, fatty acids and glycerol; the macromolecules become part of the structure of the body.

You should know:

- the need for digestion
- the role of enzymes, with specific reference to various types
- the structure and function of the digestive tract.

You should be able to:

- explain why digestion of large food molecules is essential and how it depends on enzymes
- state an amylase, a protease and a lipase, and give the source, substrate, products and optimum pH conditions for each one
- draw and label a diagram of the digestive system
- outline the function of the stomach, small intestine and large intestine
- explain the structure and function of intestinal villi.

Examples

1. What is the main function of the large intestine?
   A. Absorption of water
   B. Digestion of fats and proteins
   C. Absorption of nutrients
   D. Recycling of digestive enzymes

   Digestive enzymes are found in the small intestine, where some have come from the pancreas and others are embedded in the surface cells of the microvilli. Collectively, the enzymes complete the breakdown of disaccharides, fats and peptides. Villi, which line the walls of the small intestine, then absorb the products of digestion. Digestion of fats and proteins is the main function of the small intestine and the stomach, so B is not correct. Absorption of nutrients occurs in the small intestine, so C is not correct. Digestive enzymes are broken down, so choice D is also incorrect. This leaves A as the correct answer. After receiving the material that could not be digested or absorbed by the small intestine, the large intestine absorbs water (via osmosis) and electrolytes (via active transport), leaving behind the feces for eventual release.

2. Outline the need for enzymes in the digestive system. *[2]*

   Key points to be made are that enzymes are biological catalysts; that they speed up the rate of digestive reactions at body temperatures; that they break down large nutrient molecules; and that the nutrient molecules are reduced to a small enough size for entry into the bloodstream.

Be prepared

- Use the term "absorption" when explaining the functioning of intestinal villi. Do not confuse it with the process of assimilation, which can occur only after digested nutrients have entered the bloodstream and eventually body cells.
- When drawing a diagram of the digestive system, show that the digestive tract is a continuous tube to which other organs are joined by smaller tubes or ducts.
- You should make sure that your answers never imply that evacuation of feces could be related to excretion, as these are unrelated processes.

The transport system

Useful terminology for this section:

- **myogenic**—self-excitable property of heart muscle tissue.

You should know:

- the anatomy of the heart and the role of coronary arteries
- how the heart keeps blood moving through itself
- neural and hormonal control of heart beat
- the structure and function of blood vessels
- the types of cells in the blood and the substances transported by blood.

You should be able to:

- draw and label a diagram of the heart, showing all chambers, associated blood vessels, valves and the route of blood flow through the heart
- explain the heart pumping action with reference to valves opening and closing
- outline the control of heart muscle contraction, including its own myogenic ability, the pacemaker, nerves, the medulla of the brain and epinephrine (adrenalin)
- explain the structure and function of arteries, capillaries and veins
- state that blood is made of plasma, erythrocytes, leucocytes (phagocytes and lymphocytes) and platelets
- state that blood transports nutrients, oxygen, carbon dioxide, hormones, antibodies, urea and heat.

Examples

1. What route does blood follow to supply oxygen to heart muscle?

   A. pulmonary vein → left atrium → left ventricle → aorta → coronary artery

   B. pulmonary vein → right atrium → right ventricle → aorta → coronary artery

   C. pulmonary artery → left atrium → left ventricle → aorta → coronary artery

   D. pulmonary artery → right atrium → right ventricle → aorta → coronary artery

Here it is vital to apply three main ideas about the heart. First, atria receive blood and supply the ventricles, which pump blood. Second, the heart is a double pump in that the right side receives deoxygenated blood from the body and pumps blood to the lungs for oxygenation, whereas the left side receives fresh blood from the lungs and pumps it out to the body. Third, the heart is the most important muscle in the body, so it needs the freshest, most oxygenated blood possible. That would be supplied by coronary arteries, which are the first blood vessels to branch off the aorta. Working backwards, the aorta comes from the left ventricle, left atrium, and then the pulmonary vein. Pulmonary vein is correct since veins always carry blood towards the heart. Answer A then makes sense, because the pulmonary vein would be carrying oxygenated blood from the lungs to the heart.

2. Describe the relationship between the structure and function of blood vessels. *[6]*

The command term "describe" means detailed information should be given that relates blood vessel structure to function. The question can be broken down according to blood vessel type, since arteries, capillaries and veins are distinctly different. Arteries, carrying blood away from the heart, must have relatively thick walls that are elastic to withstand high blood pressure, and are muscular to help move the blood. Consequently, their lumens are small in proportion to overall vessel thickness. Capillaries are built for exchange, so they have thin walls, only one cell in thickness. This allows for easy diffusion of substances, either exiting or entering the blood. Capillary walls also have pores or clefts that allow ultrafiltration of substances. Finally, veins carry blood, under low pressure, back to the heart. Consequently, veins have valves to prevent backflow of blood and relatively thin walls with little elastic tissue, less muscular tissue and wider lumen than arteries.

Be prepared

- In a diagram of the heart, you should show the atria as smaller than the ventricles; the left ventricle should be larger and thicker than the right; the pulmonary artery and the aorta should be shown with semilunar valves.

Gas exchange

Required definitions for this section:

- **gas exchange**—process of exchanging oxygen and carbon dioxide in lungs and in tissues
- **ventilation**—process of bringing in fresh air to alveoli through inhalation and removing stale air from alveoli through exhalation.

You should know:

- the need for ventilation
- the anatomy and functioning of the ventilation system
- how alveoli are adapted for their role in gas exchange.

You should be able to:

- distinguish between ventilation, gas exchange and cell respiration
- draw and label the ventilation system
- describe the features of alveoli that enable gas exchange
- explain the need for ventilation, including the maintenance of concentration gradients in the alveoli
- explain the mechanism of ventilation of the lungs.

Examples

1. What feature of alveoli adapts them to efficient gas exchange?
   A. They have muscles that pump air in and out regularly.
   B. Their membranes are more permeable to gases than water.
   C. A constant blood supply flows through them.
   D. A dense network of capillaries surrounds them.

   Think of alveoli as microscopic balloons that terminate the branching airways in the lungs. Alveoli walls must be thin, one cell in thickness, to exchange oxygen for carbon dioxide from the bloodstream. Answer A is incorrect since alveolar walls are too thin for muscle tissue, and answer C is incorrect since alveoli have air in them, not blood. That leaves B and D as possibilities. B is incorrect because water movement is not part of gas exchange. That leaves answer D, which is correct because the capillary network brings blood close to the air inside the alveoli, allowing easy exchange of gases.

2. Draw a labelled diagram to show the human ventilation system. *[4]*

   Since the term "system" is used, a complete diagram would include not only the lungs but also the airways leading into the lungs and the muscles that enable inhalation and exhalation. That means the diagram should show: trachea, bronchi, bronchioles, lungs, alveoli, diaphragm and intercostal muscles. The latter muscles can be shown in between the ribs, in cross-sectional views on the left and right sides of the thoracic cavity (chest region). Remember to connect correct labels to recognizable structures.

Be prepared

- Be careful not to confuse ventilation (the exchange of air) with respiration (the cellular process).
- Include the following in your diagrams of the ventilation system: trachea, lungs, bronchi and bronchioles shown in relative proportion to each other; alveoli should be shown in an inset diagram at higher magnification. It is also advisable to include intercostal muscles, the diaphragm and abdominal muscles.
- In your explanations of the ventilation system, include how volume and pressure changes in the lungs relate to the contraction and relaxation of the internal and external intercostal muscles, the diaphragm and abdominal muscles.

Explain how the structure of a villus in the small intestine is related to its function. *[7]*

[Taken from standard level paper 2, time zone 1, May 2009]

How do I approach the question?

Since this question involves the command term "explain" and is worth 7 marks, it requires an answer that clearly relates structure to function in the villus in at least seven different ways. Many structural features help the villus absorb nutrients at the molecular level: the tiny projecting fingerlike shape gives great surface area; a wall of epithelial cells with microvilli further increases surface area; the epithelial wall thickness of only one cell minimizes the distance that nutrients must travel from the lumen of the small intestine to the interior of the villus, so passage is rapid; protein channels in the epithelial cell membranes facilitate diffusion of some nutrients; protein pumps in the epithelial cell membranes actively transport some nutrients; many mitochondria in the epithelial cells provide ATP for the active transport; the dense capillary network inside the villus is close to the epithelium, allowing rapid absorption of nutrients into the bloodstream; and the interior lacteal absorbs fats into the lymphatic system. The digestive role of the villus epithelial cells must also be included: embedded surface enzymes in the microvilli complete digestion of disaccharides and peptides. Note that, in each of the examples just given, structure is immediately related to function. This is highly advised in such a question. Though not called for, an annotated diagram can support many of these ideas.

Which areas of the syllabus is this question taken from?

- Explain how the structure of the villus is related to its role (6.1.7)

This answer achieved 2/7

1 This connects the structural feature with its purpose.

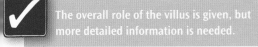

1 Should be "lacteal" not "channel".

2 Only the capillaries transport molecules to the bloodstream; lacteals transport fats to the lymphatic system.

The overall role of the villus is given, but more detailed information is needed.

The function of the small intestine is to absorb the molecules that have been broken down by digestion, the villus has a channel like a straw (1) which is surrounded by capillaries. The villus helps the small intestine because it increases the surface area and therefore more molecules can be absorbed. ✓ (1) The capillaries and the channels (2) inside the villus will transport the molecules directly to the bloodstream. ✓

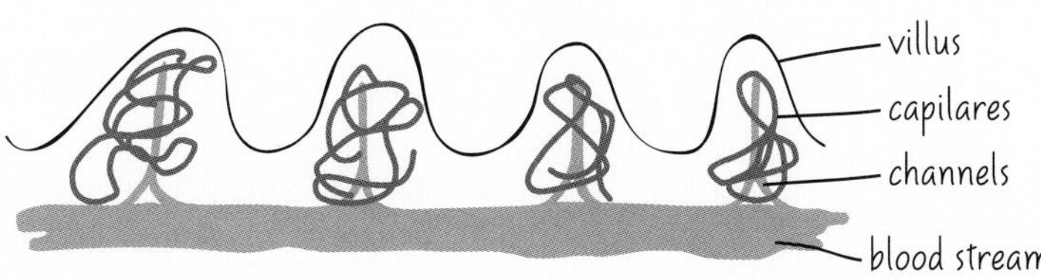

This answer achieved 5/7

1 The student has connected structural feature to function.

More details—such as embedded enzymes to complete digestion, protein channels and protein pumps to facilitate absorption, and mitochondria to provide ATP to power pumps—would complete the answer.

This answer displays many examples where structure is directly connected to function.

The structure of a villus relates to its function in many ways. First of all, its small size gives it an increased surface to volume ratio. ✔ Also, its elongated shape increases its surface area of activity. Its epithelium is one-cell thick. This allows for increased efficiency in absorption as there is a decreased distance for the material to diffuse through. ✔ 1 Its epithelium is lined with ministructures called microvilli which further increase the surface area of absorption. ✔ It has a lacteal for the absorption of fats. ✔ The capillaries in the villus serve to allow the absorbed material to diffuse into the bloodstream. ✔ Also the fact that they are very abundant further increases their efficiency for absorption by increasing the surface area of activity.

This answer achieved 7/7

1 There is an excellent build-up of ideas leading into this point.

1 Similar structures are labelled as microvilli and epithelium cells. Microvilli are on the surface of epithelium cells and their detail can only be observed under an electron microscope.

2 Depiction of the capillary is incomplete because a network is not shown around the lacteal.

The answer is slightly redundant towards the end.

This answer presents a detailed explanation showing linkage of ideas throughout.

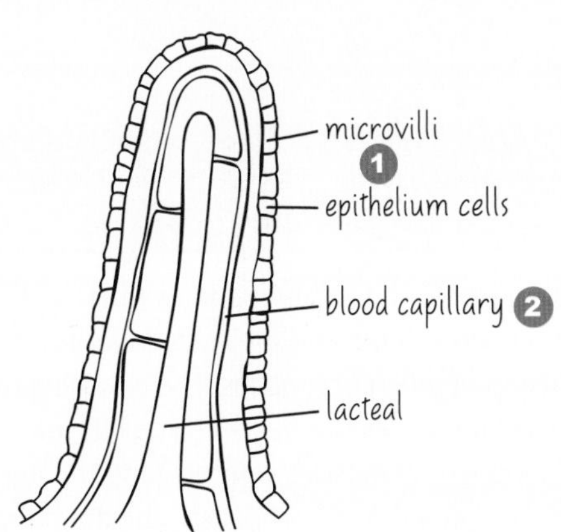

These finger-like projections that line the inside of the small intestine called villi are extremely important. Villi aid one of the functions that the small intestine does which is the absorption of nutrients. The structure of a villus is very much related to its function. Because its function is to absorb nutrients, the villi's structure plays into this. The villi is a long finger-like shape. This shape increases surface area ✓ 1 which is extremely helpful when an organism needs to absorb a lot of nutrients quickly. In addition, the microvilli on the surface of the villi serve the same function of increasing surface area for absorption. ✓ Additionally, there are many pumps ✓ located on the surface of the epithelium cells that allow for more efficient absorption of nutrients. Inside the villus are the lacteal ✓ and blood capillaries. ✓ Both of these structures also play into the villi's function. The blood capillaries are located a very short distance ✓ from the thin surface of the villus. Thus, nutrients do not have to travel a long distance to enter an organism's blood stream. Because the capillaries are located so close to the surface, villi can not only absorb nutrients rapidly but can also get nutrients into the bloodstream more quickly. Finally the lacteal serves the purpose of transporting fats out into other areas of the body ✓ where they can be used. Overall, the characteristics of a villi are closely related to its function and a major function of the entire small intestine. The villi increase surface area for more absorption, are less cell layers thick for rapid absorption, and are filled with blood capillaries for absorption straight into the blood stream. This in turn serves a purpose of the small intestine well which is the absorption of nutrients from digested food.

Explain the mechanisms involved in the ventilation of the lungs. *[8]*

[Taken from standard level paper 2, time zone 2, May 2009]

How do I approach the question?

This question is asking how air can enter and exit the lungs. The inverse relationship between gas volume and pressure underlies the event, which depends on the action of different muscles. The opposing roles of the external and internal intercostal muscles can be confusing but must be mentioned. When the external intercostal muscles contract (and the internal intercostals are relaxed), the rib cage is raised. Simultaneously, the diaphragm contracts to a flattened shape. These changes allow expansion of the lungs to greater volume, thereby lowering air pressure within the lungs. Outside air, under higher pressure, then immediately enters the lungs during inhalation. Conversely, when the inner intercostal muscles contract (and the external intercostals are relaxed), the rib cage is pulled down to press on the lungs. Simultaneously, the diaphragm has relaxed and is pushed up into a dome shape by abdominal muscles that have contracted. The resulting pressure on the lungs lowers their volume but raises the air pressure within. This high-pressure situation is relieved when air exits the lungs during exhalation. These points explain how ventilation happens; in other words, a mechanism is given.

Which areas of the syllabus is this question taken from?

- Explanation of the mechanism of ventilation of the lungs (6.4.5)

This answer achieved 3/8

1, 2 The action of the diaphragm and how its shape can be affected by abdominal muscles is well developed.

1 The comments on alveoli are not quite relevant in that ventilation involves the gross movement of air into and out of the lungs rather than the microscopic exchange of gases between alveoli and capillaries.

An overall understanding of the mechanics of ventilation is evident.

There are many mechanisms involved in the ventilation of lungs. To begin there are the external costal muscles. When these contract it causes the lungs to expand and allows one to breathe in. ✓ There are also intercostal muscles and when these contract the lungs are pushed in and one breathes out. There is also the diaphragm which is involved in ventilation. When the diaphragm contracts it flattens out which causes one to breathe in. ✓ 1 There are also abdominal muscles which, when they contract, push the diaphragm back into a dome shape and causes a person to breathe out. ✓ 2 In lungs the gas exchange is very important during ventilation. There are structures called alveoli which pass oxygen into the blood through diffusion which is then passed on to tissues. Carbon dioxide, however, is passed from tissues to blood to the alveoli. This means oxygen is inhaled while carbon dioxide is exhaled. The moist lining, thin walls, and dense capillary networks that surround the alveoli all make this process easier. 1

This answer achieved 6/8

1, 2 This is the correct use of the terms.

1, 2 The trachea volume does not change.

Ventilation means air flowing in and out of the body. When we inhale, external intercostal muscles ✓ ❶ and the diaphragm ✓ contract. The abdominal muscles relax at the same time. Contraction of diaphragm increases the volume of trachea ❶ allowing air to move faster ✓ to the lungs. When the diaphragm contracts it is flat. When we inhale internal intercostal muscles relax. ✓ ❷

Once the air arrives in the lungs, alveoli in the end of bronchioles are used to diffuse air into the blood and the rest of the body.

When we exhale, internal intercostal muscles contract. ✓ Diaphragm becomes dome shaped ✓ and abdominal muscles contract. The volume of trachea is decreased. ❷ When exhaling, carbon dioxide, CO_2, leaves the body. This process is called gas exchange; body changes carbon dioxide to oxygen.

This answer achieved 8/8

1 This is good use of scientific language.

The succinct writing style results in a quick succession of marking points to achieve the maximum.

The ventilation of lungs is the physical process of inspiration and expiration that causes air to enter and leave the lungs.

In inspiration, the diaphragm moves down by contracting ✓ and the external intercostal muscles contract, ✓ while the internal intercostal muscles relax. ✓ The ribcage expands and moves out, ✓ and the thoracic cavity ❶ expands too. Thus, the pressure within the lungs is lowered ✓ and air rushes in. ✓ This air is rich in oxygen.

In expiration, the diaphragm relaxes ✓ and moves up. The internal intercostal muscles contract, ✓ while the external ones relax. The rib cage moves in and the volume of the thoracic cavity is reduced. ✓ As a result, the pressure within the lungs increases ✓ and air is pushed out. This air is rich in CO_2.

This process is involuntary and is controlled by medulla oblongata in the brain.

Ventilation is followed by external respiration—the process of gas exchange between the alveoli in the lungs and the air there – and by internal respiration—the exchange of gases between the blood in the capillaries and the tissues.

If too much CO_2 is detected by the chemoreceptors in one's blood, then ventilation rate will increase to receive more O_2.

12. Human physiology, part 2 (Nerves, homeostasis, muscles, defence and reproduction)

Nerves

Required definitions for this section:

- **action potential**—local inversion of the electric potential between the interior and the exterior of a neuron created by a nerve impulse, followed by its restoration
- **resting potential**—electric potential between the interior and the exterior of a membrane when there is no nerve impulse propagating through it.

Useful terminology for this section:

- **neuron**—cells of the nervous system capable of carrying rapid electrical impulses
- **synapse**—location where neurons communicate between each other without being physically in contact.

You should know:

- the composition of the nervous system
- the mechanism of nerve impulse propagation
- the principles of synaptic transmission.

You should be able to:

- state that the nervous system consists of the central nervous system (CNS) and peripheral nerves, and is composed of neurons
- draw and label a diagram of the structure of a motor neuron
- state that nerve impulses are conducted from receptors to the CNS by sensory neurons, within the CNS by relay neurons, and from the CNS to effectors by motor neurons
- define resting potential and action potential (depolarization and repolarization)
- explain how a nerve impulse passes along a non-myelinated neuron
- explain the principles of synaptic transmission.

Examples

1. On the diagram of the motor neuron shown below, which label identifies a dendrite?

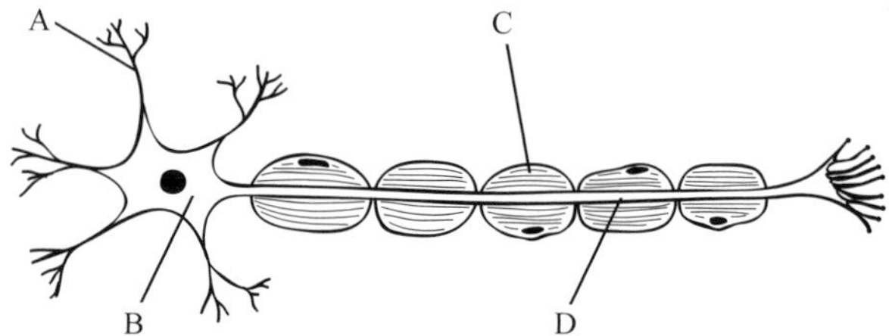

This should be an easy question based on recall to increase your overall score. The correct answer is A.

2. The graph below shows changes in membrane potential in an axon during the passage of an action potential.

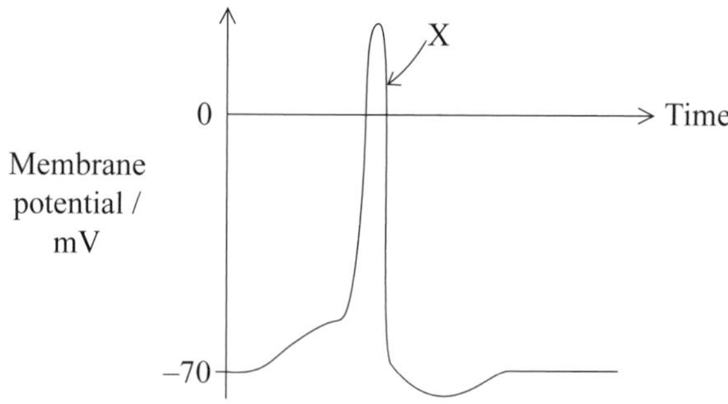

What is causing the decrease in membrane potential at point X?

A. Sodium ions entering the axon
B. Potassium ions entering the axon
C. Sodium ions leaving the axon
D. Potassium ions leaving the axon

You should remember when sodium and potassium ion channels open and close at the different phases of a graph of an action potential, and the direction of movement of the

Nerves (continued)

different ions. Potassium ion channels remain closed until the end of the rise of the action potential, but then open, letting the potassium ions out of the cell or the axon. Since the sodium ion channels have closed at the peak of the action potential, the exit of potassium ions starts to bring the inside of the axon towards a negative charge. Point X is located there, and choice D is the correct answer.

Be prepared

- A diagram of a motor neuron should show an axon at least four times the length of the cell body, as in the example above.
- You must determine the scope of questions about the nervous system and distinguish those focusing on the pathway through the CNS from others about mechanisms within a neuron and between neurons.
- Questions about nerve impulse propagation usually refer to processes happening with ion exchanges along the axon membrane, whereas questions about nerve impulse transmission between neurons refer to processes at synapses involving neurotransmitters.

Hormones and homeostasis

Useful terminology for this section:

- **hormone**—a chemical messenger that is secreted by endocrine glands into the blood, and acting on specific target cells.

You should know:

- the composition of the endocrine system
- the functioning of homeostasis
- the mechanism controlling body temperature and blood glucose
- the distinction between type I and type II diabetes.

You should be able to:

- state that the endocrine system consists of glands that release hormones that are transported in the blood
- state that homeostasis involves maintaining the internal environment between limits, including blood pH, carbon dioxide concentration, blood glucose concentration, body temperature and water balance
- explain that homeostasis involves monitoring levels of variables and correcting changes in levels by negative feedback mechanisms
- explain the control of body temperature, including the transfer of heat in blood, and the roles of the hypothalamus, sweat glands, skin arterioles and shivering
- explain the control of blood glucose concentration, including the roles of glucagon, insulin and α and β cells in the pancreas islets
- distinguish between type I and type II diabetes.

Examples

1. The diagram below represents the homeostatic control of body temperature.

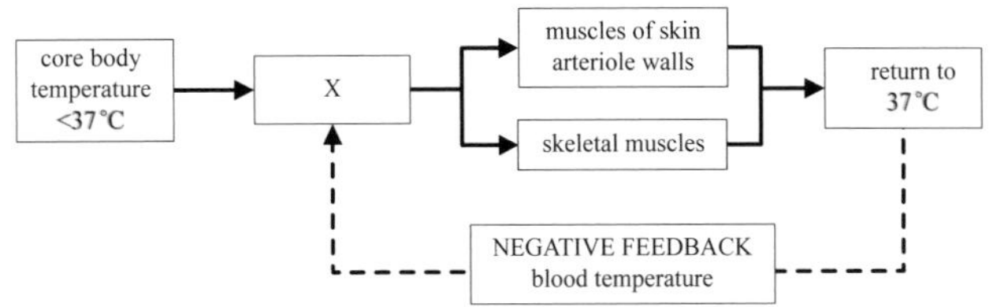

What does the part labelled X represent?

A. Heart
B. Kidney
C. Pituitary
D. Hypothalamus

You should be familiar with diagrams showing a sequence of events. In this one, part X is the structure that will sense the body temperature and will act on muscles. Muscles are effectors, and only the central nervous system can send nerve impulses towards them. The correct answer is therefore the only proposed part of the nervous system, the hypothalamus (choice D).

2. Which one of the following structures releases glucagon?

A. α cells of the pancreas
B. β cells of the pancreas
C. Liver cells
D. Hypothalamus

Hormones and homeostasis (continued)

Glucagon is produced by the alpha (α) cells of the pancreas, whereas insulin is produced by the beta (β) cells of the pancreas. This is a question of memory, and you can perhaps remember that glucagon comes before insulin in alphabetical order (thus "α"), or contains the letter "a", for alpha. Both glucagon and insulin have an action on liver cells, but liver cells do not release them. It is the pancreas, not the hypothalamus, that releases these hormones, and choice A is therefore the only valid answer.

Be prepared

- Insulin and glucagon are hormones and therefore have an action on metabolism. Your style of writing should take this into account to avoid suggesting that they would convert glucose directly into glycogen, and vice versa, as enzymes would do through reactions of condensation or hydrolysis.
- There are many homeostatic mechanisms in the body, but for the purpose of exams the range of questions is limited to the control of body temperature and glucose level. When explaining the control of body temperature, statements should mention that blood flow to capillary beds in skin can be shut down because of constriction of arteriole circular muscles at the entrances to the beds.
- You should also remember that arterioles can contract to reduce blood flow, but that capillaries cannot because they do not possess a muscular layer.

Defence against infectious disease

Required definitions for this section:

- **antibody**—a protein that recognizes and binds to the surface of an antigen to stimulate destruction of the antigen
- **antigen**—a foreign substance that stimulates the production of antibodies; foreign substances can be a wide range of substances, such as cell walls of pathogenic bacteria or fungi, and protein coats of pathogenic viruses
- **pathogen**—an organism or virus that causes a disease.

Useful terminology for this section:

- **AIDS** (an acronym for "**a**cquired **i**mmuno**d**eficiency **s**yndrome")—a group of symptoms found together (syndrome), including a low number of certain lymphocytes weight loss and little resistance to a variety of diseases (such as those caused by viruses, bacteria, fungi and protozoa)
- **antibiotic**—chemicals produced by microorganisms that kill or control the growth of other microorganisms
- **HIV** (an acronym for "**h**uman **i**mmunodeficiency **v**irus")—a virus that infects and destroys a type of lymphocyte that is necessary for antibody production in a functioning immune system; HIV causes AIDS
- **immune response**—production of specific defence proteins called antibodies in response to some intruding substance or molecule that the body recognizes as foreign.

You should know:

- the definition of a pathogen
- non-specific protection against pathogens
- antigens, antibodies and how antibodies are produced
- the limitations of antibiotics
- the danger of HIV and the AIDS dilemma.

You should be able to:

- outline the role of the skin and mucous membranes in defending against pathogens
- outline how phagocytic leucocytes ingest pathogens in the blood and body tissues
- distinguish between antigens and antibodies
- explain antibody production
- outline how HIV destroys the immune system
- discuss the cause, transmission and social implications of AIDS.

Example

1. Using a table, compare human skin and mucous membranes as barriers against pathogens. *[6]*

This question is unique because it requests the use of a table to make comparisons. The table should have two separate columns: one labelled "human skin" and the other labelled "mucous

Defence against infectious disease (continued)

membrane." Each line of the table should be a similarity or difference for the same particular property. Here is an abbreviated example.

Human skin	Mucous membrane
mechanical barrier	mechanical barrier
thick/tough/strong	thin/soft/weak
surface cells usually dead	surface cells usually alive
surface usually dry	surface usually moist

Each line of comparison (referring to the same property) would be worth one mark. Some other properties that could be compared are biochemical defences, with specific examples, and whether or not cilia are present.

2. Which term describes a molecule capable of triggering an immune response?
   A. Antibody
   B. Antigen
   C. Pathogen
   D. Antibiotic

To answer this question correctly, you must know about an immune response. By definition, an immune response is the production of proteins called antibodies. These defend against foreign substances called antigens (particles as large as pollen or as small as viruses). Thus, one can immediately identify answer A as false, and B as true. C is false because pathogens are types of antigens that cause disease. Finally, D is false since antibiotics (produced by microorganisms) are types of chemicals used to fight against pathogenic bacteria.

Be prepared

- You must know why antibiotics do not work against viruses in humans. Remember that bacteria have their own metabolic pathways, whereas viruses do not. Since antibiotics primarily interrupt the metabolic process of translation in bacteria, they can kill bacteria.
- When viruses are active in humans, they depend on the metabolic reactions of their human host cells. Antibiotics cannot recognize which cells are infected by viruses and, even if that were possible, we could not remain alive by interrupting the metabolic pathways of our own cells (in effect killing them) in order to eliminate dangerous viruses lurking within them.

Reproduction

Useful terminology for this section:

- ***in vitro* fertilization (IVF)**—fertilization using harvested eggs placed in a laboratory dish to which harvested sperm has been added.

You should know:

- the anatomy of the adult male and female reproductive systems
- the role of hormones in the menstrual cycle
- graphical trending of hormone levels during the menstrual cycle
- three roles of testosterone in males
- about *in vitro* fertilization and associated ethical issues.

You should be able to:

- draw and label diagrams of the adult male and female reproductive systems
- outline the role of hormones in the menstrual cycle, including FSH, LH, estrogen and progesterone
- annotate a graph showing hormone levels in the menstrual cycle
- outline the process of *in vitro* fertilization
- discuss the ethical issues associated with *in vitro* fertilization.

Examples

1. Explain the roles of estrogen in regulating change at puberty in young women. *[3]*

First, it can be said that estrogen production is increased during puberty. Higher levels of this hormone stimulate the appearance of female secondary sex characteristics. Specific examples of secondary sex characteristics should then be given, such as menstruation, growth in size of the vagina and uterus, increasing size of breasts, and growth of pubic and armpit hair.

Reproduction (continued)

2. What is the correct sequence of events in IVF?

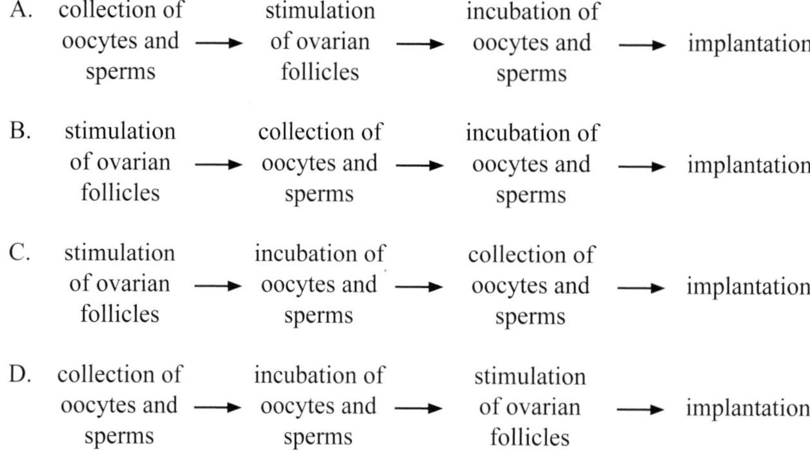

A. collection of oocytes and sperms → stimulation of ovarian follicles → incubation of oocytes and sperms → implantation

B. stimulation of ovarian follicles → collection of oocytes and sperms → incubation of oocytes and sperms → implantation

C. stimulation of ovarian follicles → incubation of oocytes and sperms → collection of oocytes and sperms → implantation

D. collection of oocytes and sperms → incubation of oocytes and sperms → stimulation of ovarian follicles → implantation

Before any oocytes can be collected, there must be stimulation of ovarian follicles to release the oocytes. Thus, choices A and D can be eliminated. Since incubation cannot occur before oocytes and sperms have been collected, answer C can also be eliminated. This leaves answer B as the logical choice, because fertilization of oocytes by sperms occurs during incubation, and that must precede implantation.

Be prepared

- Show accurate relative positions and proportional sizes of organs in your drawings of the adult male and female reproductive systems. Be sure to include the bladder and the urethra. Histological details are not expected.

Draw a labelled diagram of the adult male reproductive system. *[5]*

[Taken from standard level paper 2, time zone 1, May 2009]

How do I approach the question?

In the male reproductive system, as in most body systems, it is absolutely important to show the connections joining the organs/structures of the system, as well as their correct positioning and accurate proportional sizes. Too often, drawings simply show suspended organs without any physical relation to one another, except for their relative locations. For this diagram, the following should be shown: testes/testis enclosed by scrotum; epididymis joined to testes; sperm duct/vas deferens leading away from epididymis; bladder with urethra leading towards prostate gland; prostate gland; seminal vesicle with duct leading to prostate; penis with erectile tissue inside. Drawings of outstanding quality will also show two junctions inside the prostate. One is where a duct from the seminal vesicles joins the sperm duct, near the inside edge of the prostate, and the second is where the sperm duct joins the urethra, towards the centre of the prostate. The urethra will also be shown extending throughout the length of the penis. For this drawing, it is probably easier to show a side view rather than a front view.

Which areas of the syllabus is this question taken from?

- Diagrams of reproductive systems (6.6.1)

This answer achieved 1/5

There are many glaring shortcomings in this diagram. The sperm duct is not labelled. The epididymis appears inside the testis. An unknown duct appears to go directly from the bladder to the testis.

The labels did show some awareness of what structures should be included, but most of these are positioned incorrectly.

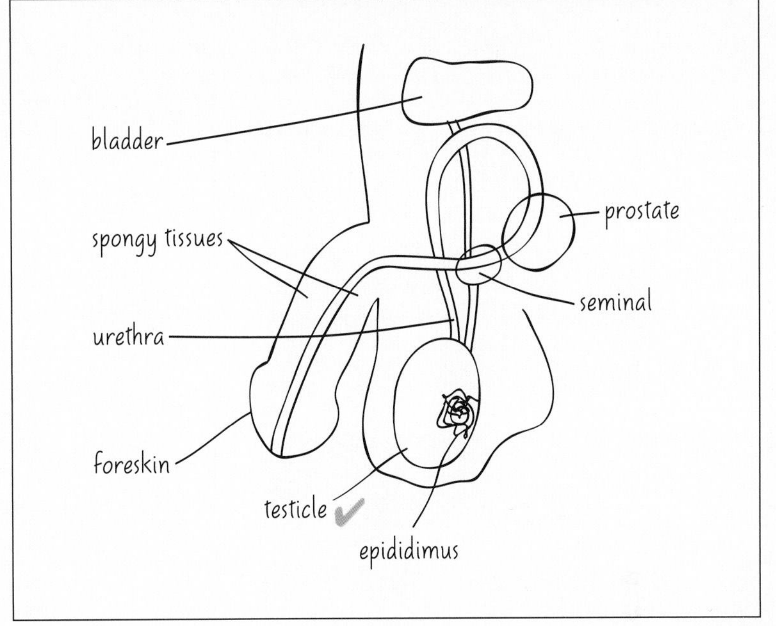

This answer achieved 3/5

1 The epididymis is well drawn, although it appears as a side view in an overall front-view diagram.

Important structures are missing, for example, penis, bladder and sperm duct. The relationship of the prostate to the urethra is unclear. A front view is depicted less often in books and is probably more difficult to draw.

Labels clearly lead to structures.

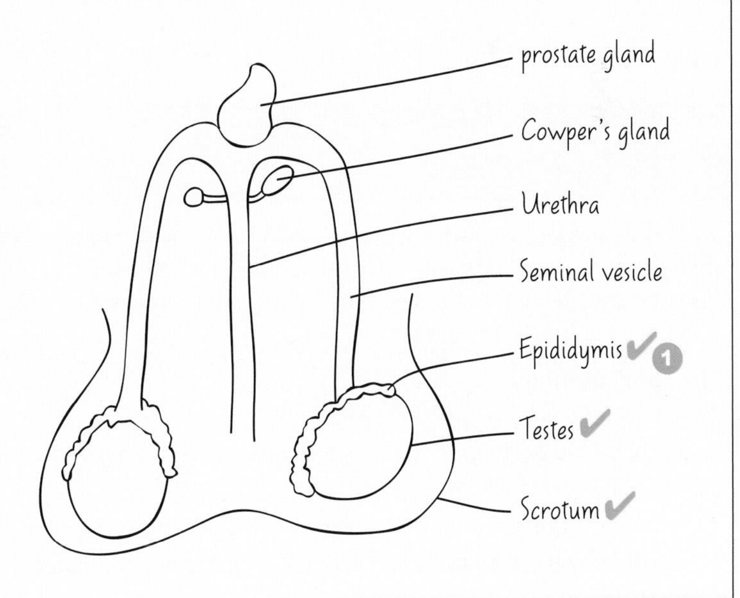

This answer achieved 5/5

An excellent understanding of ducts is shown, especially in the vicinity of the prostate gland.

1 The sperm duct is not labelled.
2 The duct from the seminal vesicle to the sperm duct is not drawn clearly and should in fact be shorter and closer to the prostate.
3 The epididymis is unlabelled and ambiguous.

The relative proportions/sizes of structures could be improved.

Except for a few missing labels, the drawing is very comprehensive.

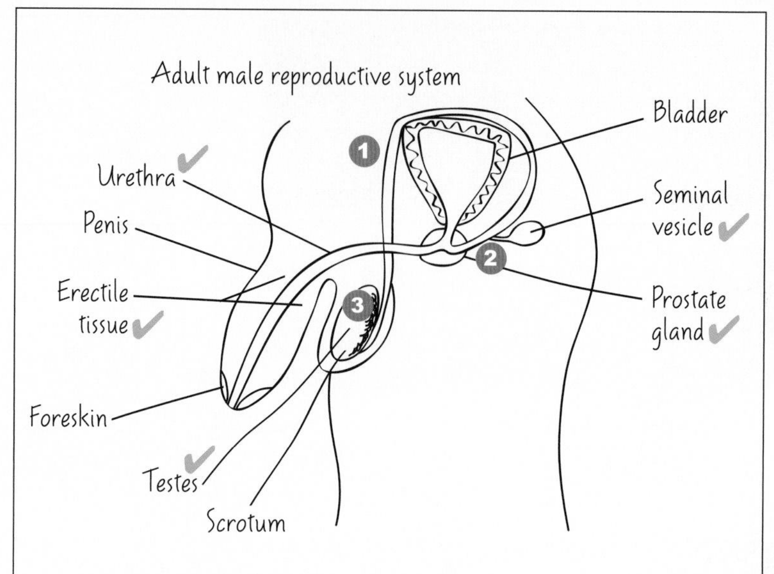

Discuss the ethical issues associated with IVF. *[6]*

[Taken from standard level paper 2, time zone 1, May 2009]

How do I approach the question?

The key in answering any discussion question is to provide as many opposing points of view ("pro" and "con" ideas) about the issue as possible. If the answer only presents one side of the argument, maximum credit can never be achieved. Here is a sampling of "pro" and "con" ideas.

The "pro" position:

- chance for infertile couples to have children
- decision to have children is a conscious one reflecting couple's acceptance of parental responsibilities
- genetic screening of embryos could decrease suffering from genetic diseases
- spare embryos can be safely stored for future pregnancies or for use in stem cell research.

The "con" position:

- IVF is expensive and might not be equally accessible
- success rate is low, so stress is placed on the couple
- infertility might be passed on to children
- unnatural process or objections based on religion or culture.

Which areas of the syllabus is this question taken from?

- Ethical issues associated with IVF (6.6.6)

This answer achieved 2/6

1 This marking point went under the category of "Accept any other reasonable answer".

No "pro" arguments are given.

For some, IVF brings up ethical issues. One issue raised by the topic of IVF is that some of the eggs taken from a woman are not used. Some people see this as "killing" ✓ 1 the potential children that could possibly be formed from these eggs. Another issue raised by IVF is the belief that fertilizing or implanting an egg in an unnatural ✓ way is changing or messing with some divine plan, or the will of a higher power.

This answer achieved 4/6

The answer attempts to provide a balanced discussion by including one "pro" idea among the many "con" ideas.

More ideas supporting IVF should be given.

The student was able to elaborate on the idea of what to do with extra embryos.

Ethical issues with IVF include controversies over how it bypasses natural reproduction. ✓ By allowing two people who could not otherwise reproduce to have babies, ✓ it may facilitate the passing on of infertility. ✓ It also involves the harvesting of gametes from the female, which may be dangerous to her health.

After fertilization, there are generally extra embryos, and what to do with these often creates controversies. ✓ For example, the throwing away of the embryos or the freezing of the embryos is controversial, especially when the frozen embryos are labelled for later use in embryonic stem cell research which is particularly controversial amongst religious and political groups. The discard of these embryos also creates controversy amongst the same groups because it raises the question of throwing away life.

This answer achieved 6/6

The last portion describing IVF and its use would have been more effective as introductory ideas for the essay.

This is a very well-organized answer, showing a balance of "pro" and "con" ideas and some depth of insight.

Arguments for IVF include that embryos with possible genetic problems can be screened ✓ before implanting, couples who couldn't previously have kids can now have a family, ✓ and eggs can be frozen to be implanted later. ✓ Arguments against IVF includes an increased risk of multiple births ✓ and problems associated with multiple births. Frozen eggs provide legal problems ✓ in cases like divorce, it bypasses nature's way of preventing reproductive problems from passing on to the next generation and parents can be selective in the features they want their child to have. ✓ In-vitro fertilization involves fertilizing eggs outside of the womb and implanting the embryos inside of the woman's uterus. IVF is useful for couples where either a man suffers from impotence or the woman has fallopian tubes which are blocked.

Components of the human diet

Required definitions for this section:

- ***cis* unsaturated fatty acid**—unsaturated fatty acid where hydrogen atoms are bonded to carbon atoms on the same side of a double bond
- **minerals**—inorganic nutrients, in an ionic form, that are required in small amounts
- **nutrient**—a chemical substance found in foods that is used in the human body
- ***trans* unsaturated fatty acid**—unsaturated fatty acid where hydrogen atoms are bonded to carbon atoms on opposite sides of a double bond
- **vitamins**—organic molecules/compounds required in the diet.

Useful terminology for this section:

- **omega-3 fatty acid**—fatty acid that has the first double bond of the hydrocarbon between the third and fourth carbon atoms counting from the end opposite the COOH group
- **rebound malnutrition**—deficiency of nutrients resulting from a high rate of excretion that continues after a change of intake from high to normal.

You should know:

- the essential nutrients for the human body and why some amino acids are non-essential
- the consequences of protein deficiency from malnutrition or from phenylketonuria (PKU)
- the structural variation of fatty acids and the health consequences of diets rich in certain types
- the differences between minerals and vitamins
- why vitamin C, vitamin D and fibre are important in human diets, and sources for vitamin D.

You should be able to:

- outline the consequences of protein deficiency malnutrition
- explain the causes and consequences of PKU and the importance of early diagnosis
- outline the structural variation of saturated fatty acids, *cis* and *trans* unsaturated fatty acids, and mono- and polyunsaturated fatty acids
- evaluate how a dietary excess of any of these fatty acids can affect health
- distinguish between minerals and vitamins regarding their chemical nature
- outline two methods to determine the recommended daily intake of vitamin C
- discuss the amount of vitamin C needed to protect adults from scurvy, upper respiratory infection and rebound malnutrition
- discuss the need for vitamin D from sunlight compared to the risk of contracting malignant melanoma
- explain how artificial dietary supplementation prevents malnutrition, using iodine as an example.

Examples

1. State **one** food source rich in vitamin D. *[1]*

   Different choices are possible. Any of the following would suffice: dairy products, (oily) fish, egg yolk and liver.

2. Explain how a special diet can reduce the consequences of phenylketonuria (PKU). *[4]*

   Begin by stating that PKU is a genetic disease in which a gene mutation prevents production of an enzyme called tyrosine hydroxylase. Without this enzyme, phenylalanine cannot be converted to tyrosine, so phenylalanine builds up in the bloodstream. This excess eventually causes dangerous consequences, such as brain damage, mental retardation, tremors, seizures, eczema and skin rashes. These problems can be avoided if, early in life, a diet is followed that has little or no phenylalanine or is low in proteins. The diet should include fruit, grain, vegetables and special formula milk.

Energy in human diets

Required definitions for this section:

- **normal weight**—BMI of 18.5–24.9 kg m^{-2}
- **obese**—BMI of 30.0 kg m^{-2} and above
- **overweight**—BMI of 25.0–29.9 kg m^{-2}
- **underweight**—BMI below 18.5 kg m^{-2}.

Useful terminology for this section:

- **body mass index (BMI)**—BMI $= \frac{\text{mass in kg}}{(\text{height in m})^2}$
- **clinical obesity**—a medical condition when extreme excess body fat can reduce life expectancy and cause other health problems.

You should know:

- the energy content of carbohydrate, fat and protein
- the staple energy sources for different ethnic groups
- the health consequences of diets rich in carbohydrates, fats or proteins
- how the brain controls appetite
- the formula for BMI and its application
- reasons for the increasing rates of clinical obesity and the consequences of anorexia nervosa.

You should be able to:

- compare the energy content per 100 g of carbohydrate, fat and protein
- compare the energy sources in the diets of different ethnic groups
- explain the possible health consequences of diets with excess carbohydrates, fats or proteins
- outline the function of the appetite control centre in the brain
- calculate BMI values and recognize their meaning in relation to an individual's weight status
- outline reasons for the increasing rates of clinical obesity and the consequences of anorexia nervosa.

Examples

1. Outline how appetite is controlled. *[3]*

A control centre in the hypothalamus portion of the brain regulates our desire for food. Whenever we have eaten enough food, various hormones (related to high blood glucose, presence of food in the small intestine or greater stored fat) are released into the bloodstream. The hormones stimulate the control centre to make us feel satiated or full of food, ending our desire to eat.

2. Describe the consequences of anorexia nervosa. *[2]*

Giving specific consequences would be best. There are many possibilities: body weight well below normal; loss of muscle mass, causing physical weakness; ending of menstrual cycle, causing infertility; low blood pressure and heart rate, causing poor circulation; brittle thinner hair; dry skin that bruises easily; and emotional problems revealed by unusual behaviour. Decisions to exercise excessively, to take laxatives or diuretics, and to induce self-vomiting would all be symptomatic of psychological problems associated with anorexia nervosa. Also, individuals with anorexia nervosa sometimes become anxious about family and friends.

Be prepared

- The availability of cheap high-energy foods, large portion sizes, increasing use of vehicles for transport, and change from active to sedentary occupations/life styles are all factors that can increase obesity.

Special issues in human nutrition

Required definitions for this section:

- **artificial milk**—milk containing glucose, bovine proteins, plant-derived fatty acids and no antibodies
- **human milk**—milk containing lactose, human proteins, human butterfat and antibodies.

Useful terminology for this section:

- **food miles**—a measure of how far a food item has been transported from its site of production to its site of consumption.

You should know:

- the composition of human milk and artificial milk, and the benefits of breast feeding
- causes and symptoms of type II diabetes and appropriate dietary advice for type II patients
- ethical issues regarding the eating of animal products
- how reducing dietary cholesterol can reduce heart disease
- the concept of food miles as a factor in making food choices.

You should be able to:

- distinguish between human milk and artificial milk for feeding babies
- discuss the benefits of breast feeding
- outline the symptoms of type II diabetes
- explain the dietary advice that type II patients should follow
- discuss the ethical issues concerning the eating of animal products
- evaluate the benefits of reducing cholesterol to minimize the risk of coronary heart disease
- discuss the concept of food miles and why consumers should choose foods with lower food miles.

Examples

1. List **two** symptoms in a patient with type II diabetes. *[2]*

Type II diabetes has many symptoms. The list includes: excessive thirst, high blood glucose levels (hyperglycemia), glucose in urine, frequent urination, slow-healing sores, swollen gums, blurred vision and erectile dysfunction.

2. Discuss whether consumers should choose foods to minimize food miles. *[3]*

Again, in order for an answer to achieve excellence, both "pros" and "cons" of the issue must be considered.

Reasons to choose foods that minimize food miles could include:

- less consumption of fossil fuel per food unit, whether by truck or airplane
- reduced air pollution and road traffic
- fresher food
- support of local growers and the local economy.

Reasons not to choose foods that minimize food miles could include:

- less choice of food, especially more exotic food
- less access to out-of-season foods
- reduced international trade that limits growth of export markets in developing countries
- foods that minimize miles may actually be more expensive.

Be prepared

- To discuss the ethics of eating animal products such as honey, eggs, milk and meat as examples of animal products.

A study investigated the dietary patterns in Chinese rural and urban populations between 1982 and 1992. The overall energy intake did not change significantly over this period.

Food source	Energy contribution by food source / percentage of total energy intake			
	1982		1992	
	Urban	Rural	Urban	Rural
Cereals	65	75	57	72
Tubers	2	9	2	4
Legumes	3	3	2	2
Animal	12	4	15	6
Energy from fat	25	14	28	19

[*Source*: Yusuf, S et al. 2001. "Global burden of cardiovascular diseases: Part 1: General considerations, the epidemiologic transition, risk factors, and impact of urbanization". *Circulation*. Vol 104, number 22. Pp 2746–2753. Copyright Wolters Kluwer. Reprinted with permission.]

(a) Identify the food source that changed least in percentage of total energy intake between 1982 and 1992 in rural Chinese populations. *[1]*

(b) Compare the changes in energy contribution by food source between urban and rural populations over the ten-year period. *[3]*

(c) Suggest possible health concerns that could result from the changing nature of the Chinese diet over this period. *[3]*

[Taken from standard level paper 3, time zone 1, May 2009]

How do I approach the question?

(a) This part of the question, for one mark, is meant to ease you into the topic. Under the "Food source" column, study the various foods to see how they compare in the 1982 and 1992 columns for "Rural" Chinese. Look for the most similar pair of numbers, which would represent minimal change in percentage of total energy intake.

(b) This part of the question is quite broad because it requires studying all the data, that is, all food sources, both urban and rural populations, and both the 1982 and 1992 columns of numbers. Furthermore, since the command term is "compare", it means that you should look for similarities **and** differences in the energy contribution of food sources throughout the data. There are many more similarities than differences to be found.

(c) Again, using all the data, you must first recognize the changes that have occurred in energy intake and then consider how they will impact health. The changes for both urban and rural populations primarily show declines in cereals, tubers and legumes but an increase in energy from animals and fat. This means that diets with more cholesterol and less fibre result in higher cholesterol levels, more heart disease, more stroke and increased bowel cancer.

Which areas of the syllabus is this question taken from?

- Health consequences of diets rich in different fatty acids (A.1.7)
- Importance of fibre (A.1.14)
- Reasons for increasing rates of obesity (A.2.7)
- Reduction of dietary cholesterol (A.3.6)

This answer achieved 4/7

1 There is accurate recognition about what dietary changes have occurred.

1 Specific reasons should be given about why the diet is unhealthy.
2 Measles is not caused by a deficiency in nutrients. It is a contagious viral disease.

Maximum credit was obtained in parts (a) and (b).

(a) The food source that changed least was legumes. ✔

(b) The urban populations consume more animal and energy from fat while the rural populations consume more cereals, tubers, and legumes. The energy intake from all food sources was only slightly higher in urban areas. Over the ten year period both rural and urban populations began consuming more animal ✔ and more energy from fat ✔ while consuming less cereal, ✔ legumes, and tubers.

(c) The overall consumption of animal is known to be unhealthy (1) and when accompanied by a lowering amount of cereal and vegetable consumption is a set up for nutrient deficiencies (1) like measles. (2)

This answer achieved 6/7

1. The information about cholesterol should mention higher blood levels of cholesterol.
2. The paragraph contains some irrelevant information that does not address health concerns related to the dietary changes noted. More appropriate information would have included atherosclerosis, stroke or bowel cancer.

Maximum credit was obtained in parts (a) and (b). There was a good analysis of the data and application of the information it contained.

(a) The food source that changed the least in percentage of total energy intake between 1982 and 1992 in rural Chinese populations was legumes. ✔

(b) The energy contributed by most food sources including cereals, tubers and legumes, over the ten-year period in both urban and rural population has either stayed the same or decreased. ✔ The two food sources that do not follow this—either staying the same or decreasing—are animal food sources ✔ and energy from fat. ✔ These two food sources have increased quite significantly over the ten-year period.

(c) The possible health concern that could result from the changing nature of the Chinese diet over the ten-year period is the risk of coronary heart disease. ✔ The risk of coronary heart disease is correlated with a steroid known as cholesterol, which is essential in membranes. Cholesterol is present in animal products. ❶ Another possible health concern that could result from the changing nature of the Chinese diet over the ten-year period is the risk of obesity. ✔ This is due to the increased energy contribution from fats.

A possible benefit that could result from the changing nature of the Chinese diet over this period is the lesser chance of having protein deficiency malnutrition which causes the swelling of the abdomen due to the inability of the plasma protein to absorb fluid from the tissues into the capillaries; but due to the increased food energy from animals, protein deficiency malnutrition shouldn't be a concern. ❷

Zinc (Zn) is an important dietary nutrient. More than 200 enzymes that are dependent on zinc have been identified. One consequence of zinc deficiency is suppression of appetite, due to reduced sensitivity to tastes and smells. A recent study compared the presence of zinc in tissue and fluid samples collected from 15 patients with anorexia nervosa to that from 15 control patients. The results are shown in the graph below.

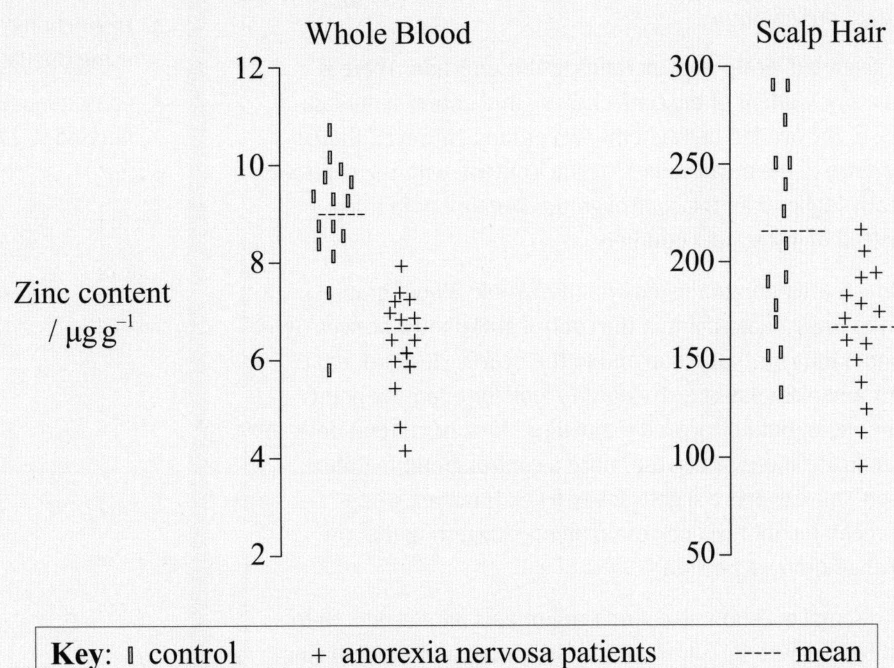

Key: ▯ control + anorexia nervosa patients ----- mean

[*Source*: Tuormaa, TE. 1995. "Adverse effects of zinc deficiency: A review from the literature". *Journal of Orthomolecular Medicine*. Vol 10, numbers 3 and 4. Pp 149–164. Fig 1.]

(a) Compare the zinc content of scalp hair of the control group with that of the anorexia nervosa group. *[2]*

(b) Discuss whether whole blood zinc content of 6 $\mu g\, g^{-1}$ would indicate that a person has anorexia nervosa. *[2]*

(c) Discuss whether dietary zinc supplementation would be an effective treatment for anorexia nervosa. *[2]*

(d) Zinc is a mineral. Distinguish between a mineral and a vitamin. *[1]*

(e) State the body mass index (BMI) below which a person is considered to be underweight. *[1]*

[Taken from standard level paper 3, November 2009]

How do I approach the question?

First, before looking at any of the questions, read the stem of the question carefully and study the data. Make sure you understand what is being said through words and data presentation.

(a) Two graphs are given but "Scalp Hair" pertains to this question. There is a similarity in the zinc content of the control group and anorexia nervosa patients because of the overlap between the two groups. However, there is an obvious difference in the mean values for zinc content, with the control group being clearly higher than the control group. Remember to make comparisons instead of just stating numbers.

(b) This question directs attention to the data on the "Whole Blood" graph. The position of the $6\,\mu g\,g^{-1}$ data point in the control group must be seen relative to the other data for that group and to the nearby cluster of data points for anorexia nervosa patients. Try hard to look for opposing points of view. For example, it could be argued that $6\,\mu g\,g^{-1}$ for zinc content does not necessarily indicate anorexia nervosa, since a control group member (a "normal person") already has a slightly lower level. However, $6\,\mu g\,g^{-1}$ is almost at the mean for anorexia nervosa patients, suggesting that the person probably has anorexia nervosa.

(c) This is another "discuss" question, so counter-arguments are needed. Here the stem should be reviewed for any important information. Note that zinc deficiency suppresses appetite in anorexia patients. Since both graphs show that the anorexia nervosa patients have lower mean values for zinc levels than the control group, it could be argued that the anorexic patients do not get enough dietary zinc, so their appetites are suppressed and they become anorexic. Therefore, dietary zinc supplements would be effective in raising zinc levels and stimulating appetite. On the contrary, it could be said that lower zinc is an effect of anorexia nervosa and not the cause. Perhaps there are other causes, such as psychological disorders.

(d) Definitions for these terms must be known before any difference can be given.

(e) Make sure to give the units for BMI.

Which areas of the syllabus is this question taken from?

- The distinction between minerals and vitamins (A.1.8)
- Distinctions of weight categories using the BMI (A.2.6)
- Consequences of anorexia nervosa (A.2.8)

This answer achieved 3/8

1 This is correct, although most vitamins are not synthesized in the body.

1 The IB marking policy is one of positive scoring. This means that accuracy will be credited even if other information is wrong, but not contradictory. No response means that the student has no chance for credit.

2 The answer does not address the question. The value given in the question is not used in relation to the graphs.

3 This is not true—carbon atoms can exist in inorganic ions such as the bicarbonate ion.

4 The units are missing. The 18.5 should include $kg\,m^{-2}$.

Some valid reasoning is shown in (b).

(a) [no response] 1

(b) It does not necessarily indicate that a person has anorexia nervosa, for anorexia nervosa is a state where a person is underweighted and sees herself in the mirror and sees herself fat. This produces them to stop eating **voluntarily** for they want to be "skinny". On the contrary, the lack of zinc produces a suppression of appetite for it reduces the sensitivity to smell and tastes. 2

(c) No, because people that suffer from anorexia nervosa stop eating because they see themselves as people with underweight so they stop eating although they might be hungry, to be more slim. If zinc supplements are introduced, this would increase their appetite ✓ which would, but not necessarily, lead to eating and then vomit what they just ate. Anorexia nervosa is a state of the mind ✓ where the girl or boy stops eating voluntarily although they might be hungry, to be slim.

(d) A mineral is inorganic. Therefore it does not contain carbon atoms 3 thus it cannot be synthesized in the body. These are found in the form of ions. On the other hand a vitamin is organic. ✓ 1 Therefore, it has carbon atoms and thus it can be synthesized by the body through other nutrients. These are found in the form of compounds.

(e) 18.5 4

This answer achieved 4/8

1 The student gains the mark when referring to the mean value of zinc content, but this is mentioned only in the second sentence.

1 The two sentences are basically just the converse of each other. Since they refer to the same idea, only one mark can be awarded.

2 This passage needs a counter-argument.

3 The irrelevant comments on enzymes muddle the answer. There is no mention of what is inorganic or organic, and no reference to size differences.

Part (c) shows two important opposing arguments.

(a) The control group had a higher zinc content in scalp hair than the anorexia nervosa group. The anorexia nervosa group had a lower mean zinc content ✓ 1 in scalp hair than the control group. 1

(b) Yes, whole blood zinc content of 6 $\mu g\, g^{-1}$ would be a strong indicator that a patient had anorexia nervosa. ✓ The data shows that the average anorexia nervosa patient has 6.7 $\mu g\, g^{-1}$ and the control patients had an average of 8.9 $\mu g\, g^{-1}$. So, zinc content below 6.7 $\mu g\, g^{-1}$ would indicate strongly that this person has anorexia nervosa. 2

(c) This supplementation would be effective in raising the patient's average whole blood zinc content, ✓ however, this doe not solve the psychological problems of an anorexia nervosa patient. ✓ A zinc supplementation will do little to solve the root of the patient's problem which is of psychological origin. The patient will still suffer from the anorexia nervosa, but one of the symptoms (low zinc) will be resolved. Many times, people think that solving a symptom will solve the problem, however, in this case, it does not.

(d) Enzymes can depend on minerals, like zinc, while vitamins are not used by enzymes. 3

(e) This often depends on the individual's height, however, an "average" person below 15–10 BMI should be considered underweight.

This answer achieved 7/8

1 The student gained the comparison mark with the word "less", but no mark would have been gained if only numerical values had been stated.

1 There is some confusion here. It would be better to write that all elements are inorganic and some elements, when occurring as ions, can be minerals.

2 This is incorrect. A mineral can contain carbon as in the bicarbonate ion.

3 Units are needed.

Answers (a), (b), (c) and (d) all properly respond to their respective command terms with thorough information provided. Overall, this was an excellent answer, showing good data analysis and comprehensive reasoning.

(a) The anorexia nervosa group had less zinc ✔ content of scalp hair (the mean was ≈170 $\mu g\, g^{-1}$) while in the control the mean was ≈220 $\mu g\, g^{-1}$. ❶ However, for some people (8) the results were the same, even if they were from different groups. ✔

(b) It could indicate anorexia nervosa as seen in the graph, this would make sense as that person would not be eating the necessary amounts ✔ $\mu g\, g^{-1}$ of variety of food needed. However, someone might have a blood ✔ zinc content of 6 $\mu g\, g^{-1}$ and not have anorexia nervosa. This would be caused from a zinc deficient diet that could be caused by economic reasons.

(c) Zinc supplementation would be effective in reversing the suppression of appetite, ✔ that has been the result of zinc deficiency. However, anorexia nervosa is more complex and would need psychological help. ✔ Even though they would feel appetite that doesn't mean they would eat and therefore not have anorexia nervosa. Anorexia nervosa is not a deficiency disease, so solving the zinc deficiency would not be a completely effective treatment.

(d) A mineral is an inorganic element ❶ found for example in rocks. It does not contain carbon ❷ and it is found in an ionic form. On the other hand, vitamins are compounds which contain carbon and are organic. ✔

(e) For someone to be underweight they should have a body mass index below 18.5. ❸

14. Option B: Physiology of exercise

Muscles and movement

Useful terminology for this section:

- **bone**—a rigid structure of the skeleton on which a force is applied to create movement
- **ligament**—a tissue that joins or wraps the structures of an articulation
- **muscle**—a tissue able to contract and apply a force to move a body structure, usually a bone
- **nerve**—a group of axons wrapped together, bringing stimuli for muscles to contract
- **sarco-**—a prefix used in terms related to muscle structure (for example, **sarco**mere, the functional molecular unit of a muscle fibre)
- **tendon**—connective tissue that joins muscle to bone.

You should know:

- role of structures involved in human movement
- structure and function of the human elbow joint
- structure of the human knee and hip joints
- cellular and molecular structure of striated muscle fibres
- molecular mechanism of muscle contraction.

You should be able to:

- state the roles of bones, ligaments, muscles, tendons and nerves in human movement
- label a diagram of the human elbow joint
- outline the functions of the labelled structures of the human elbow joint
- compare the movements of the hip joint and the knee joint
- describe the structure of striated muscle fibres
- draw and label a diagram to show the structure of a sarcomere
- explain how skeletal muscle contracts
- analyse electron micrographs to find the state of contraction of muscle fibres.

Examples

1. Label the parts of the sarcomere indicated below. *[2]*

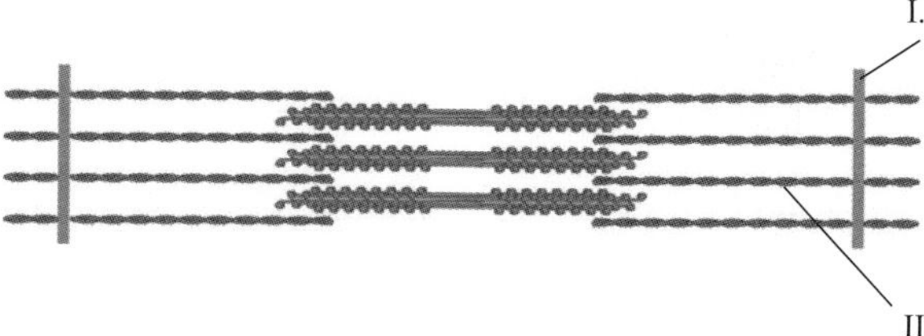

This question draws on the recognition of sarcomere parts as seen on an electron micrograph. A Z line (I) determines the limits of each sarcomere, and actin filaments (II) are attached to it. The three thick structures in the middle are myosin filaments with heads, and their width span corresponds to a dark band, whereas actin filaments of two adjacent sarcomeres separated by a Z line, in the areas where they are not overlapped by myosin, correspond to a light band.

2. Draw a labelled diagram of a human elbow joint. *[4]*

You should include the following in a diagram of the human elbow joint: cartilage, synovial fluid, joint capsule, radius, ulna, humerus, and biceps and triceps as antagonistic muscles. A variation on the same theme would be to label a provided diagram.

Be prepared

- You can make a table to compare the movements of the hip joint and the knee joint, showing similarities and differences.

Muscles and movement (continued)

- Include myofibrils with light and dark bands, mitochondria, the sarcoplasmic reticulum, nuclei and the sarcolemma in your description of a striated muscle fibre.
- Your diagram showing the structure of a sarcomere should include Z lines, actin filaments, myosin filaments with heads, and the resultant light and dark bands.
- Incorporate the release of calcium ions from the sarcoplasmic reticulum, the formation of cross-bridges, the sliding of actin and myosin filaments, and the use of ATP to break cross-bridges and re-set myosin heads in your explanation of muscle contraction. Details of the roles of troponin and tropomyosin are unnecessary and may not provide extra marks.
- You can qualify the contraction of muscle fibres as fully relaxed, slightly contracted, moderately contracted and fully contracted, if you are asked to analyse their state of contraction.
- Use appropriate terminology when writing about movement: muscles, nerves, tendons and so on have specific functions, and these terms cannot be interchanged.
- There are basically three categories of muscles in the body: striated or skeletal, smooth and cardiac. It is a good practice to specify "striated" when referring to body movement.

Training and the pulmonary system

Required definitions for this section:

- **tidal volume**—volume of air taken in or out with each inhalation or exhalation
- **total lung capacity**—volume of air in the lungs after a maximum inhalation
- **ventilation rate**—number of inhalations or exhalations per minute
- **vital capacity**—maximum volume of air that can be exhaled after a maximum inhalation.

You should know:

- terminology relating to lung performance during inhalation and exhalation
- how lung performance changes during exercise
- the effects of training on the pulmonary system.

You should be able to:

- explain the need for increased tidal volume and ventilation rate during exercise
- outline the effects of training on the pulmonary system, including changes in ventilation rate at rest, maximum ventilation rate and vital capacity.

Examples

1. Define the terms:
   (i) vital capacity
   (ii) tidal volume

   (i) Vital capacity is the maximum volume of air that can be exhaled after a maximum inhalation.
   (ii) Tidal volume is the normal volume of air taken in or out with each inhalation or exhalation, that is, in a single breath.

2. Outline the effects of training on the pulmonary system. *[2]*

   Exercise training will strengthen the rib muscles and diaphragm, allowing for an increased size of the thoracic cavity. This allows further expansion of the lungs, increasing tidal volume and vital capacity. More alveoli become inflated with each breath, so more oxygen is picked up per breath and more carbon dioxide exhaled. Also, the capillaries surrounding the alveoli grow significantly to carry more blood for gas exchange. All these changes raise VO_2 (volume of oxygen absorbed per minute and supplied to tissues) and VO_2 max (maximum rate at which oxygen can be absorbed per minute and supplied to the tissues). Finally, the ventilation rate at rest decreases from about 14 to 12 breaths per minute (bpm).

Be prepared

- Use the term ventilation rate rather than breathing rate.

Training and the cardiovascular system

Required definitions for this section:

- **cardiac output**—volume of blood pumped out by the heart per minute
- **heart rate**—number of contractions of the heart per minute
- **stroke volume**—volume of blood pumped out with each contraction of the heart
- **venous return**—volume of blood returning to the heart via the veins per minute.

Useful terminology for this section:

- **erythropoietin (EPO)**—a hormone that stimulates red blood cell production in bone marrow.

You should know:

- terminology relating to normal heart performance
- how heart performance changes during exercise
- the distribution of blood flow in the body when at rest and during exercise
- effects of training on heart performance
- risks and benefits of artificial interventions to improve cardiovascular fitness for sports.

You should be able to:

- explain the changes in cardiac output and venous return during exercise
- compare the distribution of blood flow at rest and during exercise
- explain the effects of training on heart rate and stroke volume at rest and during exercise
- evaluate the risks and benefits of using EPO and blood transfusions to improve performance for sports.

Example

Evaluate the risk and benefits of using EPO (erythropoietin) to improve performance in sports. *[3]*

For maximum credit in a question such as this, both risks and benefits must be considered. Here is a sample.

Benefits:

EPO induces production of red blood cells

more red blood cells are produced, so more oxygen is available to muscles

increased performance gives advantage over athletes who do not use EPO.

Risks:

too many red blood cells can damage capillaries, and increase risk of blood clotting, heart failure or stroke

only limited information on long-term effects is available

use of EPO can disqualify or exclude athletes from competition.

Be prepared

- Remember that changes in cardiac output during exercise involve lower blood pH, which causes impulses to be sent from the brain to the pacemaker in the heart.

Exercise and respiration

Required definitions for this section:

- **VO_2**—the volume of oxygen absorbed by the body per minute and supplied to tissues
- **VO_2 max**—the maximum rate at which oxygen can be absorbed by the body and supplied to tissues.

Useful terminology for this section:

- **creatine phosphate**—a dietary supplement that phosphorylates ADP to ATP during very short periods of intense exercise
- **oxygen debt**—amount of oxygen that is required to eliminate the lactate present in the body due to an anaerobic exercise.

You should know:

- the terms VO_2 and VO_2 max
- the roles of glycogen and myoglobin in muscle fibres and how muscle fibres produce ATP during exercise
- the effectiveness of creatine phosphate as a performance enhancer
- how intensity of exercise affects VO_2 and the amount of carbohydrate and fat used in respiration.

You should be able to:

- outline the roles of glycogen and myoglobin in muscle fibres
- outline how muscle fibres produce ATP during exercise of varying intensity and duration

Exercise and respiration (continued)

- evaluate the effectiveness of creatine phosphate in enhancing performance
- outline how intensity of exercise affects VO_2 and the changes in the amounts of carbohydrate and fat used in respiration
- outline how oxygen debt develops and how it is repaid.

Examples

1. Outline the role of myoglobin in muscle fibres. *[2]*

As an oxygen binding protein, myoglobin is found primarily in muscle tissue, where it stores oxygen. During periods of high energy demand, myoglobin will release its oxygen to oxygen-starved muscle tissue. This will allow more time for muscles to continue aerobic respiration. Afterwards, when oxygen is freely available, it can be bound by myoglobin, once again, for storage and future use.

2. Evaluate the effectiveness of dietary supplements containing creatine phosphate in enhancing performance in sports. *[4]*

"Evaluate" here means to assess the implications and limitations of creatine phosphate use. Creatine phosphate supplements are taken by body-builders to increase their muscle mass (critics say the mass increase is due to water retention). Higher muscle mass means more power for short-duration explosive events such as sprinting, weight lifting and shot putt. In these high-intensity **anaerobic** activities, reports have indicated improved performance after use of creatine phosphate. In contrast, creatine phosphate supplements are not suitable for enhancing **aerobic** exercise performance, such as long-distance running or swimming. Also, there are concerns that creatine phosphate use may suppress the body's natural ability to synthesize it. (This is significant because the body uses creatine phosphate to produce ATP.) Creatine phosphate is not recommended for adolescents or people with kidney disease, and there is an upper limit to the amount that the body can store.

Fitness and training

Required definition for this section:

- **fitness**—the physical condition of the body that allows it to perform exercise of a particular type.

You should know:

- what is meant by fitness
- the effects of exercise intensity on fast and slow muscle fibres
- some effects of performance-enhancing substances.

You should be able to:

- define fitness
- discuss speed and stamina as a measurement of fitness
- distinguish between fast and slow muscle fibres
- distinguish between the effects of moderate-intensity and high-intensity exercise on fast and slow muscle fibres
- discuss the ethics of using performance-enhancing substances.

Example

Distinguish between fast muscle fibres and slow muscle fibres. *[2]*

Slow muscle fibres	Fast muscle fibres
contract slowly	contract rapidly
resist fatigue	fatigue easily
high myoglobin	low myoglobin
aerobic metabolism	anaerobic metabolism
very good blood supply	moderate blood supply
high stamina	low stamina
more mitochondria	less mitochondria

A table such as the one given above is an efficient method to show the many differences.

Be prepared

- You must include the effects of anabolic steroids in a discussion on the ethics of performance-enhancing substances.
- EPO (erythropoietin) and creatine phosphate are performance-enhancing substances mentioned in another section for this option and can be used as examples when discussing ethical issues about performance.

Injuries

You should know:

- the need for warm-up routines
- a description of common muscle and joint injuries.

You should be able to:

- discuss the need for warm-up routines
- describe injuries to muscles and joints, including sprains, torn muscles, torn ligaments, dislocation of joints and intervertebral disc damage.

Examples

1. Discuss the need for warm-up routines. *[3]*

Although this question is worth only 3 marks, a discussion should present at least two different views. Your answer should contain at least one from each of the following "pros" and "cons".

Pros: Warm-up routines involve gentle exercise; they increase blood flow and oxygen to muscles; warm muscles are less likely to be strained.

Cons: There is limited scientific evidence for this; some people do not suffer ill effects from not warming up.

2. (i) Describe sprain injuries. *[2]*
   (ii) State **two** other types of injuries that affect muscles and joints. *[1]*

(i) Sprain injuries occur when connective tissues, ligaments and tendons, associated with a joint, become overstretched or actually tear. Sprains usually occur in joints of the ankle, wrist or finger when they are suddenly and forcefully twisted. Sometimes, in severe strains, tissues may be pulled loose from the bones to which they are attached. However, the bones remain in alignment.

(ii) Any two of the following would suffice as an answer: torn muscles, torn ligaments, dislocations and intervertebral disc damage

Be prepared

- Evidence for the purpose of discussing "pros" and "cons" of warm-up routines may be suggested by data.

Researchers have been keen to investigate whether the benefits of regular exercise go beyond physical fitness. A recent study investigated the effect of ongoing exercise on brain function:

- A control group of rats had no exercise.
- A second group of rats swam for 10 minutes each day for 15 days only.
- A third group of rats swam for 10 minutes each day for 30 days.

All three groups were tested for response to electrical stimuli during the experimental period and the response time was measured. Longer response time suggested that the rats had learnt to avoid the stimuli.

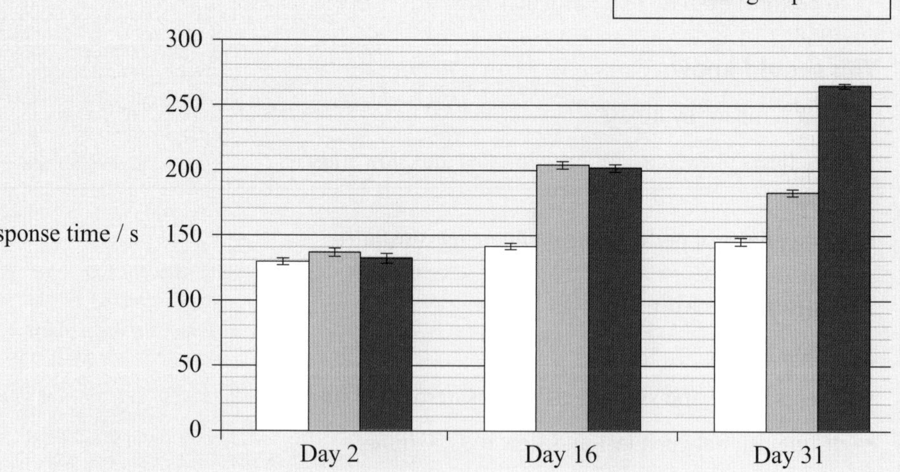

[*Source*: Parle, M, Vasudevan, M and Singh, N. 2005. "Swim every day to keep dementia away". *Journal of Sports Science and Medicine*. Vol 4. Pp 37–46. Reprinted with permission.]

(a) Identify the response time of the 15-day swimming group on day 31 of the study. *[1]*
(b) Calculate the percentage difference in response time of the control group from day 2 to day 31. *[1]*
(c) Compare the trend in response time in the 15-day and 30-day swimming groups. *[1]*
(d) Using the data above, discuss the relationship between exercise and learning. *[1]*

[Taken from standard level paper 3, time zone 1, May 2009]

How do I approach the question?

In reading the stem and looking over the data and questions, it might be difficult for you to keep track of the terminology. If so, you could design your own key of equivalent terms:

control group = no swimming = white bar

second group = 15-day group = grey bar

third group = 30-day group = black bar

(a) Locate the 15-day group in the day 31 cluster of bars. Use a square or ruler to determine the value more accurately. The response time appears to be just over 180 seconds, about 183 seconds.

(b) At day 2, the control group measures 130 seconds; at day 31, it is 145 seconds. Subtract 130 from 145 (145 – 130) for an increase of 15 seconds. Show a ratio comparing that increase against the starting value. Divide the increase by the starting value, then multiply by 100% [(15 ÷ 130) × 100%] for an answer of 11.5%. Though working is not called for, it is always a good idea to show the mathematical steps taken to calculate an answer.

(c) Here data for the 15-day group and 30-day (black bar) group must be considered, that is, day 2, day 16 and day 31. Look for similarities or differences in the data trends. It can be seen that, the longer the time spent swimming, the better the response time in group 3 (30-day = black bar); that after 16 days there is no/little difference between the two groups; that by day 31, group 3 learned most effectively; and that after day 16 the response time in group 2 drops slightly but group 3 increases further.

(d) In this question, the data for all groups and all days must be studied. Look for evidence that exercise helps learning or evidence that it does not. First of all, day 2 is a reference for comparison, since all groups started with the same ability to learn. Afterwards, by day 16, it is clear that groups 2 and 3 had better learning than the control group, which suggests that daily exercise benefits learning. By day 31, the bar for group 3, with the most swimming, obviously suggests that exercise benefits learning. Furthermore, by day 31, group 2, which had not swum for 15 days, shows a decline in learning, so the lack of swimming was reversing the learning effect. A contrary thought which could challenge the relationship between learning and exercise is that the study was based on rats and the results cannot be necessarily transferred to humans.

Which areas of the syllabus is this question taken from?

- Fitness and training (B.5)

This answer achieved 3/7

1 "Higher" and "way lower" are key words because they qualify the significance of the numbers instead of just quoting them. Marks are usually not awarded for numerical values that are just quoted.

1 It was incorrect to convert the time difference directly into percentage without first dividing by the start time.

2 Error bars are the small rotated "H" lines on top of the rectangles; the student is probably referring simply to the bars in the graph.

3 Although there is an attempt at comparing individual values, this does not answer the question of comparing trends.

4 There is no attempt at engaging in a real discussion with a range of arguments.

The student has understood the data, although showing a poor ability to compare and discuss it.

(a) The response time is about 183 s ✓

(b) Percentage difference = 0.15 or 15%

day 2 = 130 s (1.3)

day 31 = 145 s (1.45) **1**

(c) According to the error bar, **2** the response time in the 15-day and 30-day swimming groups varies depending on the number of days (2, 16, or 31). In day 2 it shows that the response time for the second group is about 138 s, which is higher **1** than 132 s, the response time for the third group. For day 16, the response time for the second group is about 204 s and is higher **1** than that of the third group response time of approximately 200 s. However, for the 31 day, different results are shown on the graph. It shows that the response time for group 2 is about 183 s, which is way lower **1** than that of the third group of about 265 s. **3**

(d) The correlation between exercise and learning are close because they go hand in hand. Based on the above, it's evident that the more one exercises the better the brain functions. ✓ Whereas, the less exercise, the less the brain functions. That's the reason why the control group of rats with no exercise had the least amount of response time. While the second group and third group with some exercise showed longer response times which shows that the rats that exercised had learnt ✓ to avoid the stimuli. **4**

This answer achieved 5/7

1 The importance of similarities or differences is mentioned.

1 It looks like "83" has been written by mistake instead of "183", but examiners cannot take this into account.

2 The difference has not been calculated.

This is a good answer addressing a comparison and a discussion correctly.

(a) 83 seconds ❶

(b) day 2 day 31
130 145

111.538% increase from day 2 to 31 in the control group of response time

145/130 = 1.11538 × 100% = 111.538% ❷

(c) From day 2 to 16 the 15-day (G2) and the 30-day (G3) groups had a similar response time ✔ with group two responding less than five seconds after group three. But once group two stopped exercising, their response time dropped from 204 seconds to 182 seconds while group 3's response time continued to lengthen. ✔ On day 31 group three had a 265 second response time in comparison to group two's 182, an 83 second difference. ❶

(d) Based on the data provided, it is hypothesized that there is a positive correlation with amount of exercise and learning. ✔ The rats that swam for 30 days continued to have a longer response time than the other two groups. ✔ It can be observed once one stops exercising their learning ability also decreases either due to exercise stimulus or lack of repetition. A lower response time correlated with lack of exercise. ✔

This answer achieved 7/7

1 It is always a good practice to include uncertainties, as in internal assessment, although there may not be a mark allocated for it.

This nice response shows a thorough understanding of the data.

(a) The brain response time was 183 seconds ±2 seconds on day 31 of the study. ✔

(b) There was increase of 15 ± 2 seconds in response time for the control group from day 2 to day 31, signifying a percentage difference of 11% ± 3%. ✔

(c) As the 15-day swimming group's response time decreased after day 16, the 30-day swimming group continued to increase ✔ (to approximately 265 s ± 2 seconds). 1 The increase was almost identical in both through day 16 (approximately 70 second increase) yet after exercising stopped the 15-day group actually decreased ✔ by about 20 seconds from day 16–31.

(d) According to the data, the more regular exercise an organism gets shows a positive direct correlation ✔ with an increase in learning. The rats with an increase in the habituality of exercise showed also significant ability to avoid ✔ the stimuli, meaning an increase in cognitive learning. Yet, if one was to stop exercising there would be a slight decrease ✔ in cognitive response (learning).

(a) Outline the relationship between the intensity of exercise and the proportions of carbohydrate and fat used in respiration. *[2]*

(b) Compare the distribution of blood flow at rest and during exercise. *[2]*

(c) Discuss speed and stamina as measures of fitness. *[3]*

[Taken from standard level paper 3, November 2009]

How do I approach the question?

(a) Subtle knowledge about energy resources and types of cellular respiration must be applied here. In terms of our bodies, invisible carbohydrates (glycogen and glucose) are usually thought of as quick energy resources, in contrast to visible fat, which can store energy for the long term.

Remember, however, they both can be used as energy resources during exercise, depending on the intensity level. With low-intensity exercise, aerobic cellular respiration actually utilizes fats in the form of soluble fatty acids rather than glucose. As the level of intensity keeps rising, a conversion from aerobic to anaerobic respiration occurs. This conversion is accompanied by a transition from fatty acid utilization to 100% glucose utilization, since glucose is the only substrate that can be used in anaerobic respiration. This can only last for a short period because lactate, a toxic waste product, is produced.

(b) When resting, the body is inactive and undergoing restoration (think of yourself after a heavy meal or when falling asleep). Blood flow to heart muscle, skeletal muscles and the skin is reduced, while blood flow to the kidneys, stomach, intestines and other abdominal organs is greatest. This distribution is reversed during exercise when movement of the body occurs (picture yourself riding a bicycle or playing a sport). Working muscles, with their metabolic needs, then require an abundant supply of blood. During rest or exercise, blood flow to the brain must remain constant since the brain integrates control of the entire body.

(c) Speed is how quickly the body or body parts move a given distance per unit of time, whereas stamina is the capacity of sustaining an effort or maintaining an exercise level. It is important to mention here that the appropriate fitness depends on the activity involved. For example, speed is important in sprinting or football, while stamina is important for rowing or cross-country skiing or running.

Which areas of the syllabus is this question taken from?

- Energy resources during respiration (B.4.5)
- Changes in blood distribution (B.3.3)
- Speed, stamina and fitness (B.5.1, B.5.3)

This answer achieved 3/7

1 The student realized that carbohydrate in the form of glucose is being used.

1 Since the mark for naming an energy source has already been given in the first sentence, the second sentence does not enable the student to gain another mark. The idea of switching from aerobic to anaerobic respiration or from fat to glucose was expected to gain other marks.

2 The focus is on rate of delivery instead of where the blood is going.

Some knowledge is displayed for each part.

(a) Short, intense sessions of exercise mean that glucose stores in carbohydrates are used. (1) ✓ Long, less strenuous sessions of exercise mean that energy stores from fat are used. (1)

(b) During exercise, the rate of blood flow increases exponentially in order to supply muscles ✓ with oxygen and nutrients and transport lactate to the liver. At rest, however, the need for such rapid transport is lower as the muscles do not require a high rate of nutrient supply and lactate deportation. (2)

(c) Speed is the maximum rate at which high intensity exercises may be carried out, while stamina refers to the maximum duration that a high intensity exercise may be carried out for. ✓ Fitness is measured by how fast and for how long a strenuous activity may be carried out at length – the longer and faster the activity is performed, the more fit the individual is.

This answer achieved 5/7

1 This is a good sequence of thought. First an idea is presented and then supported by an example.

1 There appears to be confusion, because the answer has carbohydrate and fat mixed up.

2 The answer should be more specific, and is too general as it appears.

3 This is a redundant thought.

Except for the mix-up in (a), there is ample evidence of understanding about the knowledge being tested.

(a) With the increase in exercise intensity, the proportion of carbohydrate to fat begins to move in favour of fat, while at low intensity exercises, carbohydrate is metabolized more than fat. (1)

(b) Blood is concentrated more in the lungs, heart, and skeletal muscle during exercise. ✓ While resting, blood is more concentrated at the stomach, liver and kidneys. The brain always has the same supply of blood. ✓

(c) Speed is the time taken to perform a movement. (2) Stamina is the time an exercise can be maintained for. ✓ Both are measures of fitness, however it depends on the activity. ✓ Sprinting would be measured by speed and marathons would be measured by stamina. ✓ (1) Fitness is the body's ability to perform a particular type of exercise so it depends on the activity. (3)

This answer achieved 7/7

These are solid answers. A thorough command of the information is demonstrated.

(a) At low intensity exercise aerobic respiration occurs allowing the use of carbohydrates and fat. As exercise intensity increases, the proportion of fat used is reduced ✓ until respiration is anaerobic using 100% carbohydrate. ✓

(b) During exercise blood flow to the skeletal muscles and skin increases more than at rest. ✓ The blood flow to the brain remains constant at rest and during exercise. ✓ Blood flow to the stomach however decreases during exercise when compared to blood flow at rest.

(c) Speed measures the time taken to perform a specific exercise. Stamina measures the duration an exercise can be continued. ✓ Both can be useful measures of fitness, however, some activities are better suited for one measure of fitness. ✓ For example: speed is used as measure of fitness in 100m sprints. ✓

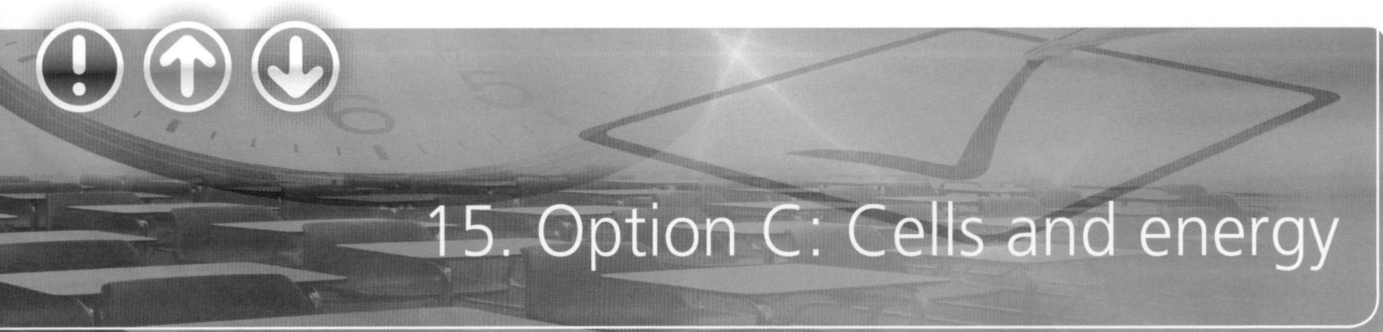

15. Option C: Cells and energy

Proteins

You should know:

- the structure and function of proteins.

You should be able to:

- explain the significance of the four levels of protein structure, including polar/non-polar amino acids where relevant
- outline the difference between fibrous and globular proteins, using two examples for each
- state four named examples of protein functions.

Examples

1. (a) Explain the primary structure of proteins and secondary structure of proteins. *[3]*

   (b) Using named examples, distinguish between fibrous protein and globular protein. *[2]*

   (a) This short explanation will contain a brief description for each of the structural levels, and the reason for each level will be limited to the nature of the bonds. The primary structure of proteins would be their sequence and number of amino acids in a polypeptide chain, because that determines all the other levels of structure in a protein. Also, the peptide bonding between the amino acids could be considered primary because it joins all the amino acids in a polypeptide chain. The secondary structure of proteins is the folding pattern, which gives an alpha-helix or beta-pleated sheet. These patterns occur because of hydrogen bonding between the carboxyl group of one amino acid and the amino group of a nearby amino acid.

   (b) For each part of your answer about one type of protein, the command term "distinguish" requires that there must be a counterpart about the other type if you want it to be a "distinction". To be able to gain all marks, you must also include named examples for each type, such as collagen for fibrous and hemoglobin for globular. Distinctive features are long shape for fibrous and rounded for globular; usually structural for fibrous and functional for globular; and usually insoluble in water for fibrous and soluble for globular.

2. The following illustration shows the three-dimensional structure of a protein.

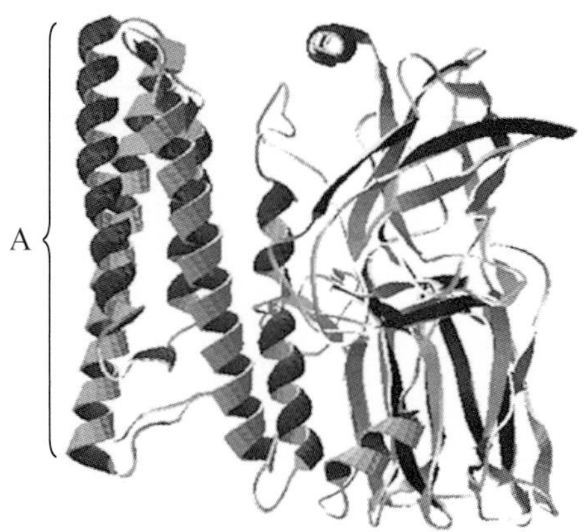

[*Source*: Soberón, M, Fernández, LE, Pérez, C, Gill, SS and Bravo, A. 2007. "Mode of action of mosquitocidal *Bacillus thuringiensis* toxins". *Toxicon*. Vol 49, number 5. Pp 597–600. Reproduced with permission.]

   (a) State the type of structure in the region marked "A".

   (b) Outline how this structure is held together.

   (c) Region A inserts into a plasma membrane. Deduce the nature of the amino acids that would be expected to be found in this region.

   (a) For this part, you must recognize that this is an α-helix of the secondary level of protein structure.

   (b) This structure is held together because of hydrogen bonds between the C–O and H–N groups of the amino acids.

Proteins (continued)

(c) To deduce the nature of the amino acids that would enable this part to insert into a plasma membrane, a link to the nature of the plasma membrane is necessary to establish that the hydrophobic non-polar tails of phospholipids of the membrane are involved. Therefore, the amino acids must be non-polar to insert them into, otherwise they would be repelled.

Be prepared

- Explaining the significance of the four levels of protein structure requires a certain depth of knowledge, as shown by the examples above.

Enzymes

Useful terminology for this section:

- **allosteric site**—location on an enzyme where an inhibitor can bind, changing the shape of the enzyme along with the active site
- **competitive inhibition**—the situation where an inhibiting molecule that is structurally similar to the substrate molecule binds to the active site, preventing substrate binding
- **non-competitive inhibition**—the situation where an inhibitor binding to an enzyme causes a conformational change in its active site, resulting in a decrease in activity.

You should know:

- the difference between competitive and non-competitive inhibition
- the roles of enzymes in controlling and inhibiting metabolic pathways.

You should be able to:

- state that metabolic pathways consist of chains and cycles of enzyme-catalysed reactions
- describe the induced fit model
- explain that enzymes are catalysts because they lower the activation energy of the reactions they catalyse
- explain the difference between competitive and non-competitive inhibition, with reference to one example of each
- explain the control of metabolic pathways by end-product inhibition, including the role of allosteric sites.

Examples

1. Explain what is meant by allosteric inhibition. *[3]*

   Since this is a short explanation, your statements should briefly link facts to consequences. You can state that allosteric inhibition is a form of non-competitive inhibition, where the inhibitor binds to a site that is not the active site; this causes a conformational change in the enzyme; this changes the active site, and the substrate can no longer bind to the active site.

2. Describe the induced fit model of enzyme action. *[2]*

   Your description must give a detailed account of what happens in this model. You can mention that enzyme and substrate join and that the enzyme active site changes shape to fit the substrate; this allows some enzymes to catalyse several substrates. It is the enzyme that changes shape to fit the substrate, not the converse. You must use the words "active site" and "substrate" correctly to be able to score marks.

Be prepared

- Properties of enzymes can be questioned indirectly in exams through other areas of the syllabus, for example, digestive enzymes (chapter 11).
- The syllabus limits non-competitive inhibition to the situation where an inhibitor molecule binds somewhere other than an enzyme's active site.
- Active site and allosteric site(s) should be clearly distinguished.

Cell respiration

You should know:

- the processes of cell respiration in relation to the structure of a mitochondrion.

You should be able to:

- state the various ways that oxidation and reduction can occur in terms of gain/loss of electrons, oxygen or hydrogen
- state the role of oxidation and reduction in cell respiration
- outline the four stages of glycolysis—phosphorylation, lysis, oxidation and ATP formation
- outline where and how phosphorylation, lysis, oxidation and ATP formation occur during glycolysis
- explain aerobic respiration, including the link reaction, Krebs cycle and the electron transport chain
- describe the role of NAD and (NADH + H^+) and oxygen in aerobic respiration
- draw and label a diagram of the structure of a mitochondrion as seen in an electron micrograph
- explain the relationship between structure and function in the mitochondrion
- explain how energy is released incrementally in the electron transport chain
- explain the mechanism of chemiosmosis and its dependence on the electron transport chain
- explain how oxidative phosphorylation and chemiosmosis are interdependent
- explain the role of ATP synthase and oxygen in cell respiration
- discuss oxidative phosphorylation in terms of the electron transport chain, oxygen as the ultimate (final) electron acceptor and the work of ATP synthase.

Examples

1. Outline the role of oxygen in cellular respiration. *[2]*

   The question asks for an outline and is worth only 2 marks, so only a brief answer should be provided. The answer should mention that it is involved in the electron transport chain, as a final electron/hydrogen acceptor, combining with H^+ (and electrons) to produce water.

2. State one difference between oxidation and reduction. *[1]*

   Any one of these three differences would be a valid answer:

   oxidation is addition of oxygen, reduction is the removal
   oxidation is removal of H^+, reduction is the addition
   oxidation is loss of electrons, reduction is gain.

Be prepared

- NAD and (NADH + H^+) in respiration must be distinguished from NADPH and (NADPH + H^+) in photosynthesis (remember "P" for photosynthesis, although it really stands for phosphate).
- When drawing a mitochondrion, use correct proportions to show and label the outer membrane, the inner membrane with folds (cristae) and the presence of enzymes, the free ribosomes and the mitochondrial DNA.

Photosynthesis

You should know:

- the relation between pigments and the structure and function of chloroplasts
- the processes of photosynthesis.

You should be able to:

- explain the relationship between the absorption spectrum and the action spectrum
- draw and label a diagram showing the structure of a chloroplast and explain the relationship with its function
- explain the mechanisms of the light-dependent reactions, including the photoactivation of photosystem II, photolysis of water, electron transport, cyclic and non-cyclic photophosphorylation, photoactivation of photosystem I, and reduction of $NADP^+$
- explain the mechanisms of the light-independent reactions, including the role of ribulose bisphosphate (RuBP), carboxylase, reduction of glycerate 3-phosphate (GP) to triose phosphate (TP), NADPH + H^+, ATP, regeneration of RuBP, and subsequent synthesis of more complex carbohydrates
- explain photophosphorylation in terms of chemiosmosis
- explain the effects of temperature, light intensity and carbon dioxide concentration in terms of limiting factors.

Example

1. State **one** enzyme involved in photosynthesis. *[1]*

   The most likely choice is ribulose bisphosphate carboxylase, which catalyses carbon fixation in the light-independent reactions. That enzyme is also called RuBP carboxylase or Rubisco. Another enzyme of great importance is ATP synthase, also called ATPase, which converts ADP to ATP in the light-dependent reactions.

2. (a) State the location in the chloroplast of the following reactions in photosynthesis. *[2]*
   (i) Light-independent reactions
   (ii) Light-dependent reactions

   (b) Explain what happens to the electrons in the light-dependent reactions of photosynthesis. *[3]*

(a) Note that light-independent reactions are listed before light-dependent reactions. This is unusual because the initial reactions of photosynthesis require the absorption of light energy, hence the term light-dependent reactions. Absorption of light energy occurs in chlorophyll pigments embedded in the thylakoid membrane, so the light-dependent reactions occur in the thylakoid membrane. Light energy is eventually converted to the chemical energy of ATP needed for the light-independent reactions. ATP is produced just outside the thylakoid membrane, in a region called the stroma. That makes ATP readily available to power the light-independent reactions that occur in the stroma.

(b) During the light-dependent reactions, electrons located in the chlorophyll pigments of the thylakoid membrane become photoactivated by absorbing photons of light. The electrons are raised to higher energy levels and then passed from one carrier molecule to the next in the thylakoid membrane. At each passage, a small amount of energy is released, which means that the electrons eventually drop back to their original energy levels before photoactivation. The released energy is bound in ATP molecules, which are produced during the light-dependent reactions.

Be prepared

- The absorption and action spectrum can be represented by graphs in exams.
- Explaining the process of photosynthesis requires a detailed account of events, in a logical or chronological order and the use of appropriate terminology.
- For light-dependent and light-independent reactions, students should understand that products of one reaction become the substrates of the other and vice-versa (ADP, ATP, NADP), and that this can become a reason for a plateau in the rate of reaction. Whereas products and substrates are renewable, light is not renewable.
- Analysis of data relating to photosynthesis may be required in exams.

The graph below shows the results of an experiment to determine the effect of salt (NaCl) concentration on photosynthesis of the freshwater green alga *Chlorella vulgaris*. The experiment attempted to determine the effect of salt concentration on the light-dependent reactions overall and separately on photosystem I and photosystem II.

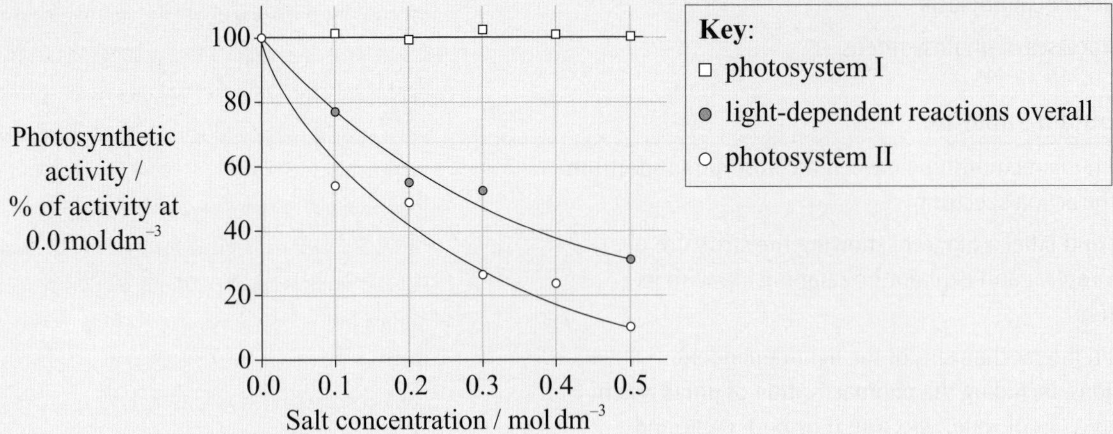

[*Source:* El-Sheekh, MM. 2004. "Inhibition of the water splitting system by sodium chloride stress in the green alga *Chlorella vulgaris*". *Brazilian Journal of Plant Physiology*. Vol 16. Pp 25–29. Copyright *Brazilian Journal of Plant Physiology*. Reproduced with permission.]

(a) Describe the effect of salt concentration on the activity of the light-dependent reactions overall. *[1]*

(b) Compare the effect of increasing salt concentration on photosystem I with the effect on photosystem II. *[1]*

(c) When salt concentration is increased, some algal cells increase their rates of cyclic photophosphorylation. Deduce the reasons for this. *[2]*

(d) Using the graph, predict the effect of high salt concentration on the growth of *Chlorella vulgaris*. Give a reason for your answer. *[2]*

[Taken from standard level paper 3, November 2009]

How do I approach the question?

Start out by carefully reading the stem and carefully analysing the graph with its key. Make sure you understand the experiment before proceeding.

(a) The key says that three lines appear in the graph. The top line, for photosystem I, appears as a straight horizontal grid line, making it somewhat subtle at first glance. However, this question is only about the middle graph line representing "light-dependent reactions overall". It is easily seen that the pattern is an inverse relationship, that is, the photosynthetic rate of light-dependent reactions goes down with increasing salt concentration.

(b) In this comparison question, the only similarity between photosystems I and II is that they are both 100% active at zero salt concentration. With increasing salt concentration, the difference between the photosystems becomes more and more pronounced, that is, photosystem I remains 100% active but photosystem II immediately starts declining in activity and continues declining throughout the experiment.

(c) Recall that, during the light-dependent reactions, sunlight can induce both photosystems to release excited electrons. In what is called non-cyclic flow of electrons, photosystem II releases excited electrons, which are accepted by photosystem I. These electrons replace other excited electrons that photosystem I has already released to reduce $NADP^+$. In the case of this experiment, increasing salt concentration keeps lowering the activity of photosystem II, yet photosystem I keeps operating at 100% activity through increasing rates of cyclic photophosphorylation. What is the reason for this? As increasing salt reduces the ability of photosystem II to supply enough electrons to photosystem I, photosystem I can replace those losses by recycling some of the excited electrons it normally releases to $NADP^+$. These electrons enter a cyclic pathway where carriers guide them back to photosystem I to maintain a steady supply, which keeps photosystem I active at 100%. Although the recycled electrons never reduce $NADP^+$, their passage during cyclic photophosphorylation results in the production of ATP.

(d) The graph shows that increasing salt lowers the overall photosynthetic rate of the light-dependent reactions. Since these reactions enable carbon fixation to occur in the light-independent reactions of photosynthesis, the overall rate of photosynthesis in *Chlorella vulgaris* will be reduced. This will mean limited growth of the algae.

Which areas of the syllabus is this question taken from?

- Light-dependent reactions (C.4.3)
- Analysis of data relating to photosynthesis (C.4.9)

This answer achieved 2/6

1. This represents a substantial gap in understanding, since glycolysis has nothing to do with photosynthesis.
2. The student has made no response, which means that there is no chance for credit.

The candidate correctly reads the graph. The comparison in (b) is complete.

(a) The higher the salt concentration is the smaller the photosynthetic activity is. ✔

(b) Increasing the salt concentration in photosystem I doesn't present a substantial change in the level of photosynthetic activity. The graph shows that it maintains the photosynthetic activity always constant. However increasing the salt concentration in photosystem II shows that the greater the concentration the smaller is the photosynthetic activity. ✔ In order for cellular respiration to take place oxygen is needed.

(c) Glycolysis takes place. ❶

(d) [No response is given.] ❷

This answer achieved 4/6

1. The comparison is complete.
2. Benefit of the doubt was given since reduced growth is implied, but not clearly stated.

1. No understanding of cyclic photophosphorylation is evident.

Good reasoning is shown in (d), that is, photosynthesis → *Chlorella* → autotroph → food production → growth.

(a) As the salt concentration increases the photosynthetic activity of the light-dependent reaction overall decreases. ✔ In an increase of 0.5 mol dm^{-3} of salt concentration the activity decreased by 70%.

(b) Photosystem II decreased by more percentage of activity than light dependent reactions. Meanwhile photosystem I remained constant, it was unaffected. ✔ ❶

(c) The algal cells might increase their rates of cyclic photophosphorylation because they have more phosphates present. The activity is increased due to the salt concentration. ❶

(d) As salt concentration decreases the photosynthetic activity ✔ this could affect the growth of *Chlorella vulgaris* as an autotroph. This means it would make less food and therefore its growth will be affected. ✔ ❷

This answer achieved 6/6

1 It is good to see the use of a mathematical term often relevant to biological relationships.

2 Some might just say "unaffected by salt concentration". This answer is superior because it includes the word "increasing".

1 This statement is incorrect.

Well-expressed accurate answers are given throughout all parts.

(a) It has a negative correlation, (1) as salt concentration increases, light-dependent reactions overall decrease. ✓

(b) The effect of increasing salt concentration on photosystem II is that it decreases its activity while it seems to have no effect on photosystem I. ✓

(c) Cyclic photophosphorylation involves photosystem I. ✓ The graph shows that photosystem I is unaffected by increasing salt concentration ✓ (2) but that photosystem II activity decreases with higher salt. This is why increasing salt concentration increases the rate of photosynthesis. (1)

(d) High salt concentration would decrease the growth of *Chlorella vulgaris* ✓ because photosystem II and light-dependent reactions overall have decreased and that would prevent an efficient photosynthesis ✓ to take place to form essential chemical substances for the plant such as glucose, starch. The fact that the *Chlorella vulgaris* grows on fresh water also might infer this.

(a) State the location of high proton concentration caused by electron transport in the mitochondrion. *[1]*

(b) Outline the role of oxygen in cellular respiration. *[2]*

(c) Explain how any two structural features of the mitochondrion are related to its function. *[2]*

[Taken from standard level paper 3, November 2009]

How do I approach the question?

This question requires specific knowledge about the structural features of mitochondria that enable it to perform aerobic cellular respiration and chemiosmosis.

(a) It should be known that every mitochondrion has two membranes, an outer membrane and an inner membrane, which is expanded by infoldings called cristae. Between these membranes is the intermembrane space, which is separated from the interior (matrix) of the mitochondria by the inner membrane. Embedded in the inner membrane is a series of electron carriers, which, during aerobic cellular respiration, transfer high-energy electrons. As each carrier in the chain accepts and then donates these electrons (originating from food), small amounts of energy are released. The energy is used to transport protons (hydrogen ions) from the matrix side, across the inner membrane, to the intermembrane space, where they accumulate, forming a proton gradient.

(b) When electrons have reached the end of the electron transport chain, they have no energy left. The final electron acceptor in the chain then passes these electrons to oxygen (on the matrix side), which also picks up a pair of protons (hydrogen ions) to form water.

(c) Various features could be explained. Here is a sampling of three examples with their purpose:
- cristae for increasing surface area of the inner membrane, which allows more electron carriers;
- small inter-membrane space for rapid build-up of concentration gradient containing transported protons;
- fluid matrix containing enzymes and other chemicals to support unique chemical reaction (Krebs cycle) of aerobic cellular respiration.

Which areas of the syllabus is this question taken from?

- Explanation of aerobic respiration (C.3.4)
- Explanation of oxidative phosphorylation (C.3.5)
- Structure and function of the mitochondrion (C.3.6)

This answer achieved 1/5

1 This should be "aerobic cellular respiration".

2 When nothing is written, there is no chance for credit.

The important role of oxygen is recognized, although it could be improved by saying "oxygen is the final electron acceptor of the electron transport chain and accepts H to form water".

(a) pili

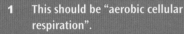

(b) In order for cellular respiration **1** to take place, oxygen is needed. Oxygen accepts electrons. ✓

(c) [No response given.] **2**

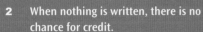

This answer achieved 2/5

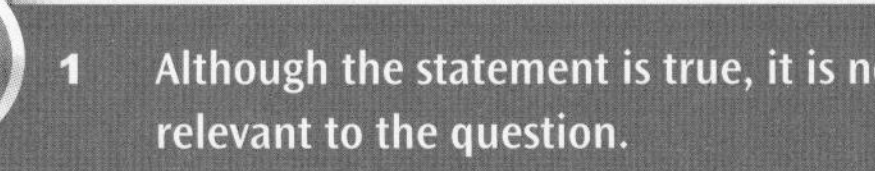

1. Although the statement is true, it is not relevant to the question.
2. The phrase "in its presence" refers to oxygen, but its role must be explained.
3. This is irrelevant information.
4. There is not quite enough for a mark. The structural feature would be the presence of chemicals in the matrix that permit the Krebs cycle to occur there.

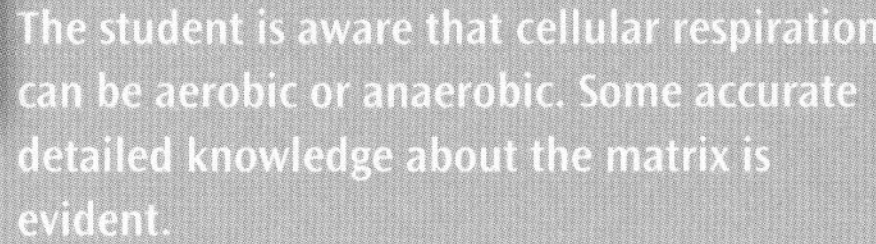

The student is aware that cellular respiration can be aerobic or anaerobic. Some accurate detailed knowledge about the matrix is evident.

(a) Between the inner and outer cell membrane of the mitochondria. ✔

(b) Cellular respiration is not dependent on oxygen. It may occur in its absence through anaerobic respiration. **1** This happens, for example in the human muscle when vigorous exercise happens, lactic acid is produced. It also happens in its presence—that is called aerobic respiration. **2**

(c) (1) Small space between inner and outer cell membrane creates a high proton concentration. ✔ This is useful for the electron transport chain.

(2) The matrix uses a big portion of the cell, it has adequate amount of ribosomes 70s. **3** The matrix is where the Krebs cycle takes place. **4**

This answer achieved 3/5

1 Very good information is given here.

1 The information about oxygen is too general. The specific role of oxygen is needed.

A good explanation appears in (c).

(a) Between the outer and inner membrane of the mitochondrion. ✔

(b) Used in oxidative decarboxylation. Differentiates between anaerobic and aerobic respiration because if oxygen is present then the pyruvate molecules can change for the production of energy in the form of ATP and the waste produce CO_2 and H_2O. **1**

(c) Cristae allow for large surface area for the electron chain reactions to take place. ✔ The fluid matrix contains large amount of enzymes needed for the Krebs cycle. ✔ **1**

16. Option D: Evolution

Origin of life on Earth

You should know:

- processes required to spontaneously originate life on Earth
- possible origins of organic compounds on Earth
- the importance of RNA and protobionts for the origin of life
- possible role of prokaryotic cells in forming the early atmosphere and eukaryotic cells.

You should be able to:

- describe the processes needed to spontaneously originate life on Earth
- outline the experiments of Urey and Miller
- discuss the conditions and locations for the synthesis of organic compounds, and state a possible alternative origin for them
- state examples of protobionts and a reason for why they may have preceded living cells
- outline two properties of RNA that may have allowed it to help life begin
- outline the possible role of prokaryotes in the formation of an oxygen-rich atmosphere
- discuss the endosymbiotic theory for the origin of eukaryotes.

Examples

1. Outline **two** pieces of evidence that support the endosymbiotic theory of the origin of eukaryotes.

   The command term "outline" requires a brief account or summary. An answer such as "70S ribosomes" is not sufficiently elaborated to provide an account. "Like prokaryotes, mitochondria and chloroplasts have 70S ribosomes within them" provides the expected amount of detail. The bold "**two**" in the question indicates that you should not outline more than two pieces of evidence, as only the first two may be marked.

2. Outline the experiments of Miller and Urey into the origin of organic compounds.

   An outline requires a brief summary or account. For this question, you should summarize the purpose, design, results and conclusions of the experiment. There is often confusion about the gases used in the experiment.

Be prepared

- The theories discussed in this part of the syllabus are explanations that can only be tested using models. A range of possible explanations exist.
- Protobionts are not cells.
- Clay provides the conditions for the reactions, rather than actually replicating molecules.

Species and speciation

Required definitions for this section:

- **allele frequency**—the commonness of an allele, as a proportion of all alleles of the gene in a population
- **gene pool**—all the genes in an interbreeding population at a certain time.

Useful terminology for this section:

- **adaptive radiation**—a form of divergent evolution when a population spreads out until new selective pressures cause the population to diversify into several species occupying different ecological niches
- **allopatric speciation**—type of speciation when members of a species that have become geographically isolated undergo changes that eventually prevent them from breeding with the rest of the species
- **convergent evolution**—type of change over time when natural selection, acting the same way in different parts of the world, causes distantly related species to evolve similar traits
- **divergent evolution**—a form of speciation when one ancestral species becomes a group of species
- **gradualism**—evolution over a long period of time when large changes take place slowly
- **polyploidy**—the presence of more than two sets of homologous chromosomes in an organism
- **punctuated equilibrium**—expression to describe short periods of rapid evolution that occur between long periods of no change
- **speciation**—formation of a new species by the splitting of an existing species
- **sympatric speciation**—type of speciation when members of a species living in the same area evolve until they can no longer breed with the rest of the population.

You should know:

- types of speciation
- types of evolution
- different ideas on the pace of evolution
- types of polymorphism.

You should be able to:

- discuss the meaning of the term "species"
- describe three examples of barriers to gene pools
- explain how polyploidy can contribute to speciation
- compare allopatric and sympatric speciation
- compare convergent and divergent evolution
- outline the process of adaptive radiation
- discuss the pace of evolution with reference to gradualism and punctuated equilibrium
- describe transient polymorphism, with an example
- describe balanced polymorphism, using the example of sickle-cell anemia.

Examples

1. Outline the pace of evolution as implied in the theory of punctuated equilibrium.

   An "outline" requires a brief account or summary. In this case, begin by defining punctuated equilibrium, and then provide a summary of the main points of the theory. These include: long periods with little change interrupted by short periods of rapid evolution, which occurs when environmental change introduces new selection pressures. An example to illustrate the theory should be included.

2. Compare allopatric and sympatric speciation.

   For "compare" questions, students are usually advised to summarize similarities as well as differences. A table could be used to organize your answer. You could also use a Venn diagram (a Venn diagram symbolically shows conceptual areas of overlap using circles). A similarity is that both allopatric and sympatric speciation involve reproductive isolation and therefore isolation of gene pools. An example to illustrate both types is recommended.

Be prepared

- Reproductive isolation is often used to describe a population that cannot breed with other members of the same species.
- Geographic isolation is not the only type of reproductive isolation.
- The definition of the words allele, chromosome and gene are often confused.
- A change in phenotype frequency does not necessarily mean a change in allele frequency.

Human evolution

Required definitions for this section:

- **cultural evolution**—changes in accumulated knowledge that occur through language, via teaching and learning, during a lifetime or over generations
- **genetic evolution**—changes in the inheritable traits of a species over generations
- **half-life**—the time needed for 50% of the radioactive atoms in a sample to decay (transform into another element).

You should know:

- the use of radioactive isotopes to date fossils
- trends in the fossil record and its incompleteness
- the major anatomical features that humans share with primates
- hominid evolution
- genetic and cultural evolution in recent human evolution.

You should be able to:

- outline the method for dating rocks and fossils using ^{14}C and ^{40}K
- deduce the age of a material using a simple decay curve for a radioisotope
- describe the anatomical features that define humans as primates
- outline the trends seen in the fossils related to hominids
- discuss the incompleteness of the fossil record
- discuss how change in diet correlates to increase in brain size during hominid evolution
- discuss the relative importance of genetic and cultural evolution in recent human evolution.

Examples

1. Explain the correlation between the change in diet and increase in brain size during hominid evolution.

   This explanation requires **both** reasons **and** mechanisms. Answers should take care when discussing cause-and-effect relationships, as the question refers to correlation. Growth in the brain requires more protein. Eating meat provides more protein.

2. (i) State **one** species in the genus *Australopithecus*. *[1]*

   (ii) For this species, state the geographical distribution and approximately how many years ago it lived. *[2]*

   Your answer should state the species accurately, provide a date within an acceptable range and indicate the area of distribution ("Africa" only is not acceptable). The species name contains the genus (starting with an upper-case letter) and the species (starting with a lower-case letter) name. Since the genus appears in the question, you can use the abbreviation "A." for it. Two species are mentioned in the syllabus: *A. afarensis*, from 3.9 to 2.9 mya (million years ago) in eastern Africa, and *A. africanus*, from 2 to 3 mya in southern Africa. *A. garhi*, from 2 to 3 mya in eastern Africa, is also an acceptable answer.

Be prepared

- *Ardipithecus ramidus*, *Australopithecus* (including *A. afarensis* and *A. africanus*) and *Homo* (including *H. habilis*, *H. erectus*, *H. neanderthalensis* and *H. sapiens*) provide fossils from which you can outline trends in human evolution.
- You should state accurately which isotope can be used to date sample material of different ages, based on their half-life.

The soapberry bug (*Jadera haematoloma*) feeds on the seeds of plants from the soapberry family (Sapindaceae). It does this by penetrating the fruit containing the seeds with the mouthparts called the proboscis. The diagrams below show sections through the fruits taken from two members of the Sapindaceae family.

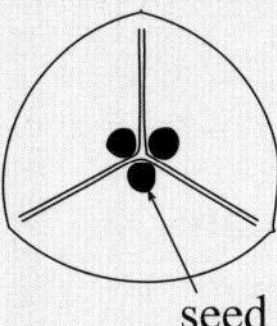

Cardiospermum corindum
(native species)

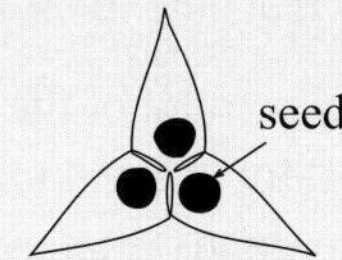

Koelreuteria elegans
(introduced species)

[*Source*: Carroll, SP and Boyd, C. 1992. "Host race radiation in the soapberry bug: Natural history, with the history. *Evolution*. Vol 46. Pp 1052–1069. Copyright © Wiley-Blackwell. Reproduced with permission.]

In Florida, *Cardiospermum corindum* is native to the area while *Koelreuteria elegans* is a species that was introduced in the 1890s and is now common in Florida. The graph below shows proboscis lengths of samples of adult female soapberry bugs in Florida between 1880 and 1980.

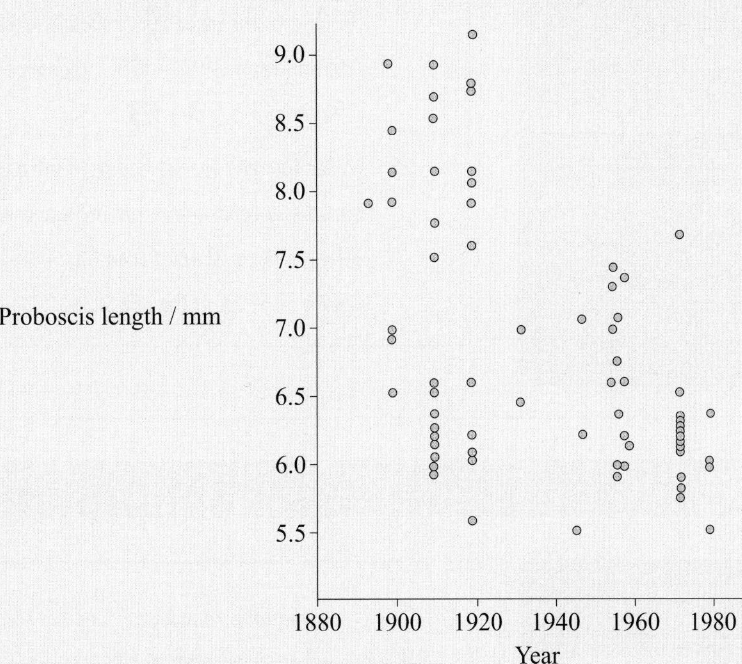

[*Source*: Carroll, SP and Boyd, C. 1992. "Host race radiation in the soapberry bug: Natural history, with the history. *Evolution*. Vol 46. Pp 1052–1069. Copyright © Wiley-Blackwell. Reproduced with permission.]

(a) Compare the fruit of the native species with that of the introduced species. *[2]*

(b) (i) Outline the trends in proboscis length in soapberry bugs shown in the graph. *[2]*

(ii) Explain how the change in proboscis length could have occurred. *[3]*

[Taken from standard level paper 3, November 2009]

How do I approach the question?

(a) For this "compare" part, be sure to draw in similarities **and** differences. In this question, you should note the difference of how much flesh in the fruit must be penetrated before reaching the seeds.

(b) An "outline" requires a brief account of the trend. The number of points allocated is usually an indicator of the number of distinct ideas students are being asked to notice in the trend—that average length is going down and that variation is decreasing.

Which areas of the syllabus is this question taken from?

- Speciation (D.2)

This answer achieved 1/7

1. Being out of 2 marks, the student should attempt to find at least one similarity and one difference. The student should use more precise terminology, such as the flesh is thicker.
2. Note that "large" does not necessarily equal "long".
3. The student is describing the trend rather than outlining the trend.
4. The student has not stated explicitly that the introduced species has become more common.
5. The student should use more specific language: the flesh is thicker, the length increases.

(a) The *C. corindum* is fatter. (1)

(b) (i) There is a general decrease in proboscis length from 1880–1920. ✔ The large (2) proboscis were between 9.2 mm and 7.5 mm. From 1960–1980, the large proboscis are between 7.5 mm and 6.5 mm. (3)

(ii) After the new species had been introduced, the new species becomes a food source for the soapberry bugs. (4) They have different sizes of fruit from the native species. (5) The proboscis length of the bugs then develops to adapt the new species.

This answer achieved 7/7

1. The student's word choices show conceptual understanding.

1. While the student earns the mark, the wording of the answer should have been more carefully chosen. Variation is a core concept, so the student might have chosen to say that the variability in length has decreased.

The reasonably careful use of language helps the student to score well on this question.

(a) The *C. corundum* is rounder ✔ and has seeds that are smaller. ✔ The *K. elegans* has larger seeds and the seeds are closer to the surface.

(b) (i) The proboscis length of the soapberry bugs decreased from 1880 to 1980. ✔ In 1880, the length had a big range but in 1980, the lengths were more concentrated. ✔ (1).

(ii) The introduced fruit, *K. elegans* has seeds closer to the outside of the fruit, ✔ which favours soapberry bugs with shorter proboscis length. This would give bugs with a shorter proboscis a higher chance of surviving and reproducing, ✔ hence passing on their genes. ✔ Bugs with a longer proboscis length have been outcompeted. (1)

(a) Describe **two** anatomical features of humans that define them as primates. *[2]*

(b) Discuss the implications of the incompleteness of the fossil record on our understanding of human evolution. *[3]*

[Taken from standard level paper 3, time zone 1, May 2009]

How do I approach the question?

(a) Be sure to describe only those features that are relevant to the classification of primates. A description requires going beyond a list to provide a brief outline of the feature.

(b) Fossilization is a rare occurrence. Most components of formerly living things tend to decompose relatively quickly following death, with only the hardest parts being likely to fossilize. In this question, you should summarize a range of perspectives on the fragmentary fossil record. The main idea in this question is that paleoanthropology is a data-poor science, with much room for uncertainty.

Which areas of the syllabus is this question taken from?

- Human evolution (D.3.4, D.3.7)

This answer achieved 2/5

1 The second part of the answer identifies features of the genus *Homo*.

2 The answer has insufficient detail about the consequences of basing theories on limited data.

(a) Two major anatomical features defining humans as primates are opposable thumbs ✓ and a relatively "flat" face, with non-protruding jaws and protruding bony brows above the eyes. **1**

(b) Due to the fact that very few organisms died "untouched", or being eaten, had their body parts scattered everywhere, it is hard to find fossils that are complete. ✓ As a result, we do not have much data to work with **2** and thus can't pinpoint exact intermediates. Without intermediates, we do not know who we evolved from and thus we don't understand human evolution quite that well yet.

This answer achieved 5/5

1 Rather than use tightly worded statements, the student elaborates, or describes fully, the requested features.

2 To ensure full marks, the student goes far beyond explaining just a few points to include a range of possible answers.

(a) 1. Their arms joined to their body in a manner which allows them to move their arms in 3 dimensions. ✓

2. Their eyes located in front of their face with overlapping fields of views to enhance vision. ✓ 1

(b) Fossil records are not complete for various reasons; this hinders our understanding of human evolution. Fossil records may be incomplete because of the following reasons:

- Decomposers break up human/pre-human remains ✓
- Environmental conditions weather fossil remains ✓
- Fossils only form when the organism is layed under sedimentary rocks
- Not all fossils have been found; many remain buried
- Some fossils found are usually incomplete, leaving humans to make inferences for the rest of the body.

These reasons all implicate our understanding of human evolution since our understanding is based on the inference of the very few fossils ✓ we have found. Since not all fossils have been found for various species, this leaves a gap in the timeline of human evolution. Also fossils found may not represent the whole population accurately. To further our understanding of human evolution, humans must continue searching for remnants of the past in the form of fossils. 2

17. Option E: Neurobiology and behaviour

Stimulus and response

Required definitions for this section:

- **reflex**—a rapid, unconscious response
- **response**—an action on the environment (internal or external) following a stimulus
- **stimulus**—a change in the environment (internal or external) that is detected by a receptor and elicits a response.

Useful terminology for this section:

- **effector**—group of cells or organ able to act on the environment following a stimulus
- **reflex arc**—simple pathway in the nervous system between a receptor, an unconscious part of the CNS and an effector.

You should know:

- the meaning of stimulus, response and reflex
- the role of different types of neurons in response to stimuli
- effects of natural selection on animal responses.

You should be able to:

- define the terms stimulus, response and reflex in the context of animal behaviour
- explain the role of receptors, sensory neurons, relay neurons, motor neurons, synapses and effectors in the response of animals to stimuli
- draw and label a diagram of a reflex arc for a pain withdrawal reflex
- explain how animal responses can be affected by natural selection.

Examples

1. Define the terms *stimulus* and *reflex*. [2]

   You can use the definitions provided at the beginning of this section. All key words are required, including "rapid" for reflex.

2. Explain the role of neurons used in the pain withdrawal reflex. [3]

   You can build your answer in three or four steps, indicating the name and function of the neurons involved in each step. It is important that you use the appropriate terminology. Sensory neurons receive the information from receptors; these transmit nervous impulses to the central nervous system (CNS); relay neurons in the CNS transmit the information from sensory neurons to motor neurons; which in turn send impulses to the effector (muscle).

Be prepared

- When questioned about reflexes, your answers should show that you have understood that reflexes only involve unconscious parts of the nervous system (for example, spinal cord and brain stem).

Perception of stimuli

You should know:

- the diversity of stimuli that can be detected by human sensory receptors
- the structure of the human eye
- the cell types in the retina in relation to light
- how visual stimuli are processed
- the structure of the ear
- how sound is perceived by the ear.

You should be able to:

- outline the diversity of stimuli that can be detected by human sensory receptors, including mechanoreceptors, chemoreceptors, thermoreceptors and photoreceptors
- label a diagram of the structure of the human eye
- annotate a diagram of the retina to show the cell types and the direction in which light moves
- compare rod and cone cells
- explain the processing of visual stimuli, including edge enhancement and contralateral processing
- label a diagram of the ear
- explain how sound is perceived by the ear, including the roles of the eardrum, bones of the middle ear, oval and round windows, and the hair cells of the cochlea.

Example

(a) Label the diagram of the retina below. *[2]*

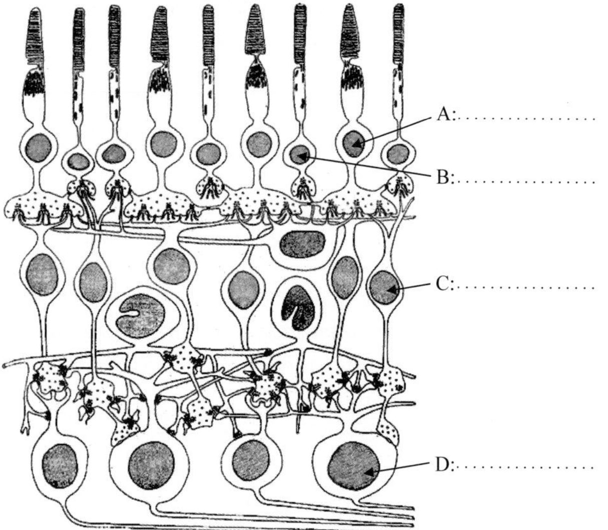

[*Source*: Dowling, JE and Boycott, BB. 1966. "Organization of the primate retina: Electron microscopy". *Proceedings of the Royal Society B*. Vol 166, number 1002. Pp 80–111. Copyright JE Dowling and BB Boycott. Reproduced with permission.]

(b) Draw an arrow on the diagram of the retina to indicate the direction in which light is moving. *[1]*

(a) This is a relatively easy question based on recall of factual information, but you must be careful not to confuse rods, which are thinner, with thicker cones. You should also pay attention to the structures indicated by the lines, as the first cell on the right is a rod, but the line for structure A points to the cell next to it, which is a cone. Since only 2 marks were allocated for this question, you can assume that two correct structures were required to gain 1 mark.

(b) The light travels through the retina and is reflected back on the choroid towards the receptor cells (rods and cones). A clear arrow going through the retina and pointing towards the top of the diagram is all that was required to gain the mark.

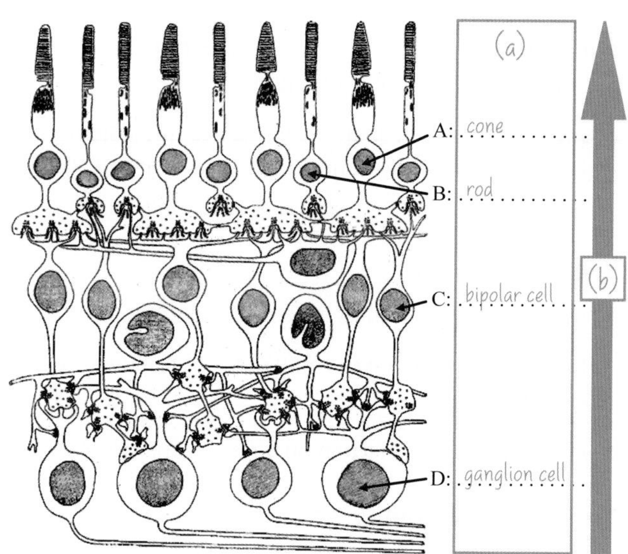

Be prepared

- The variety of questions that could be asked in exams from this section is limited. The way they could be linked to a data-based question is unpredictable, but variations on the following types of questions have appeared in past exams: very short questions about the name of sensory receptors; annotations of a diagram of either the eye, the retina or the ear; a comparison of rods and cones; and longer questions on either the processing of visual stimuli or the mechanism of sound perception.

Perception of stimuli (continued)

- Make sure that you can include or label the sclera, cornea, conjunctiva, eyelid, choroid, aqueous humour, pupil, lens, iris, vitreous humour, retina, fovea, optic nerve and blind spot in a diagram of the structure of the human eye.
- You must recognize rod and cone cells, bipolar neurons and ganglion cells out of many other types of cells in a diagram of the retina provided for labelling. You must also figure out the direction of the light through the retina, as in the above example.
- Present a comparison of rod and cone cells as a table opposing the following elements.

Rod cells	**Cone cells**
use in dim light	use in bright light
one type sensible to all visible wavelengths	three types sensible to red, blue and green light
impulses pass from a group of rod cells to a single nerve fibre	impulses pass from a single cone cell to a single nerve fibre

- You must recognize the pinna, eardrum, bones of the middle ear, oval window, round window, semicircular canals, auditory nerve and cochlea on a diagram of the ear to be able to label them.
- You must be able to recognize structures presented in other diagrams than those used in your usual textbook. Explore images on the Internet and flip/rotate them to obtain different views.

Innate and learned behaviour

Useful terminology for this section:

- **kinesis**—the movement (as opposed to growth) of an organism in response to the intensity of a stimulus, but not to its direction (for example, woodlice/pillbugs increasing their movement in all directions when placed in a dry environment)
- **taxis**—locomotion of an organism in a particular direction in response to an external stimulus (for example, movement towards light in positive phototaxis).

You should know:

- the distinction between innate and learned behaviour
- how to design experiments to investigate innate behaviour
- how to analyse some data on behaviour
- the relationship between learning and survival
- Pavlov's experiments on conditioning
- the role of inheritance and learning on development.

You should be able to:

- distinguish between innate and learned behaviour
- design experiments to investigate innate behaviour in invertebrates
- analyse data from invertebrate behaviour experiments in terms of the effect on chances of survival and reproduction
- discuss how the process of learning can improve the chance of survival
- outline Pavlov's experiments into conditioning of dogs
- outline the role of inheritance and learning in the development of birdsong in young birds.

Examples

1. Distinguish between taxis and kinesis. *[2]*

   This question only requires differences to be given. Taxis is directional (moving towards or away from a stimulus), whereas movements related to a kinesis increase with the intensity of the stimulus and are not directional. It is important that you mention the relation of kinesis with intensity.

2. Discuss how the process of learning can increase chances of survival. *[2]*

   In this very short discussion, you can only present a couple of examples where learning increases chances of survival, such as by avoiding dangerous situations or predators, enabling more successful hunting or obtaining food, or adapting behavioural strategies to changing environmental conditions.

Innate and learned behaviour (continued)

Be prepared

- Innate behaviour to be investigated on invertebrates include either a taxis or a kinesis.
- Limit yourself to invertebrates, that is, animals without a vertebral column, such as Annelida, Arthropoda and Mollusca (chapter 10), when designing an experiment about innate behaviour. (Woodlice/pillbugs are the commonly used Arthropoda to investigate a taxis or kinesis.)
- You should transfer the skills that you developed for internal assessment to answer a question asking you to design an ethical experiment on innate behaviour. Select appropriate independent, dependent and controlled variables, and describe the method used to control variables and collect a sufficient amount of relevant data; naming the organism is part of controlling the variables.
- Include the terms unconditioned stimulus, conditioned stimulus, unconditioned response and conditioned response in your answers on conditioning of dogs.

Neurotransmitters and synapses

You should know:

- the principles of postsynaptic excitation and inhibition
- the interaction between excitatory and inhibitory presynaptic neurons
- the mechanism of action of psychoactive drugs on postsynaptic transmission
- examples of excitatory and inhibitory psychoactive drugs
- the effects of tetrahydrocannabinol (THC) and cocaine on synaptic activity
- factors causing addiction.

You should be able to:

- state that some presynaptic neurons excite and others inhibit postsynaptic transmission
- explain how decision-making in the CNS can result from the interaction between the activities of excitatory and inhibitory presynaptic neurons at synapses
- explain how psychoactive drugs affect the brain and personality by either increasing or decreasing postsynaptic transmission
- list three examples of excitatory and three examples of inhibitory psychoactive drugs
- explain the effects of THC and cocaine in terms of their action at synapses in the brain
- discuss the causes of addiction, including genetic predisposition, social factors and dopamine secretion.

Examples

1. List **two** examples of inhibitory psychoactive drugs. *[1]*

Since only 1 mark is allocated for this question, you must list the names of two of these drugs. Benzodiazepines, alcohol and tetrahydrocannabinol (THC) are all inhibitory psychoactive drugs mentioned in the syllabus.

2. Explain how excitatory psychoactive drugs affect the brain. *[3]*

In this short explanation, although you could provide a definition of excitatory psychoactive drugs, you should concentrate on a few mechanisms of action for these drugs. You must use specific terminology. Examples of actions are increasing postsynaptic transmission, causing chemical dependence, producing psychomotor arousal (increased alertness) by acting like neurotransmitters or interfering with the breakdown of neurotransmitters. These actions can affect the transmission of the visual signals in the thalamus or the optical cortex.

Be prepared

- You must have understood the general principles of synaptic transmission (chapter 12) to be able to explain the effects of excitatory and inhibitory substances on synapses.
- The names that you should remember are nicotine, cocaine and amphetamines for excitatory drugs, and benzodiazepines (also known as valium and diazepam), alcohol and tetrahydrocannabinol (THC) for inhibitory drugs.
- Your explanations of the effects of psychoactive drugs must be written in terms of synaptic mechanisms.

(a) Compare rods and cones. [3]

(b) Explain the role of receptors, sensory neurons and motor neurons in the response of animals to stimuli. [3]

(c) List **four** general kinds of sensory receptor. [2]

[Taken from standard level paper 3, November 2009]

How do I approach the question?

(a) The only similarity between rods and cones is that they are both light receptors, and the main part of your answer should therefore distinguish them. A table is always the best way to present comparative or distinctive elements, providing that equivalent elements are presented next to each other. There is no problem if you present your answer as a text, but you must take care to present a counterpart for each element. Your answer could look like this, three pairs of correct elements being enough to gain full marks:

Rods	**Cones**
work better in dim light	work better in bright light
absorb all visible light wavelengths	different kinds of cones absorb different groups of wavelengths, mainly red, blue and green
are spread out through the retina	are concentrated in the centre of the retina
many rods connect to one neuron	one-to-one connection to a neuron

Do not be afraid to expand your answer and state clearly that there are three different kinds of cones, each specialized for each main colour, instead of associating them vaguely to colour vision.

(b) Your answer should state clearly what each of the three question elements actually does in creating a response to stimuli, using the correct terminology when it applies. The receptors detect stimuli. Sensory neurons send nerve impulses to the central nervous system (CNS). The CNS sends impulses to an effector, via motor neurons.

(c) Your answer should contain a complete list, since each mark is probably allocated for two elements. The following elements covered by the syllabus are expected: thermoreceptor, chemoreceptor, photoreceptor, mechanoreceptor. Baroreceptor and propioceptor are also valid answers.

Which areas of the syllabus is this question taken from?

- Comparison of rod and cone cells (E.2.4)
- Role of receptors, neurons and effectors in the response to stimuli (E.1.2)
- Human sensory receptors (E.2.1)

This answer achieved 3/8

1. It is not because they pick up black and white that they are better in dim light.
2. Using "different cones" would be preferable.
3. The CNS should be mentioned.
4. The mechanism is not specified.
5. Receptors (cells) were confused with sensory organs.

(a) Rods are very sensitive to light and therefore are best used in dim light as they pick up black and white, ❶ whereas cones ❷ pick up red, green and blue light ✔ so ❶ are best in bright light. ✔

(b) In a stimulus to reaction event, the receptors allow the animal to notice the stimuli, ✔ the sensory neurons carry the message of notice, ❸ and the motor neurons help facilitate a reaction. ❹

(c) ❺

1. light receptor (eye)
2. sound/vibration receptor (ear)
3. smell receptor (nose)
4. taste receptor (taste buds or tongue)

This answer achieved 5/8

1. This is a complete list, containing all required elements.

1. Elements are not compared to their counterpart (dim versus bright; black and white versus colour).
2. Mechanoreceptors in muscle do not sense touch.
3. CNS is missing.
4. These should have been identified as being the effectors.

(a) Rods and cones are light receptors in the eye. Rods are more sensitive to light ✔ and therefore function in dim light whereas cones are specialized to receive red, blue and green light, therefore allowing us to see colour. ❶

(b) Receptors first recognize the stimuli ✔ e.g. mechanoreceptors in muscles ❷ recognize touch. This message is then sent to sensory neurons to determine nature of stimuli ❸ and the message is sent to the motor neurons ✔ for response e.g. photoreceptors in eyes recognize bright light, the motor neurons then stimulate constriction of iris muscles. ❸❹

(c) ❶

1. chemoreceptors
2. thermoreceptors ✔
3. photoreceptors
4. mechanoreceptors ✔

This answer achieved 8/8

This is a clear and concise answer containing all the required elements.

(a) Rods function better in dim light whereas cones function better in bright light. ✔ Multiple rods attach to bipolar and ganglion cells whereas only one cone cell will attach to bipolar and ganglion cells. ✔ Rod cells have a low resolution and aren't effective at recognizing colours whereas cone cells have high resolution ✔ and can detect colours.

(b) Receptors pick up the stimuli which are then sent to the sensory neurons ✔ that send the message to the central nervous system. ✔ The CNS then sends a response which is carried by the motor neurons that cause the response in the muscles. ✔

(c) 1. chemoreceptor
2. photoreceptor ✔
3. mechanoreceptor
4. thermoreceptor ✔

Scientists studied the flight behaviour of bumblebees (*Bombus terrestris*) searching for artificial flowers of various sizes and colours. The search time is the time taken from leaving the first flower to landing on the second flower. The colour contrast is an arbitrary value which shows the colour contrast of the flowers with their green leaf-like type background. The graph below shows the search time for flowers of different colours and sizes.

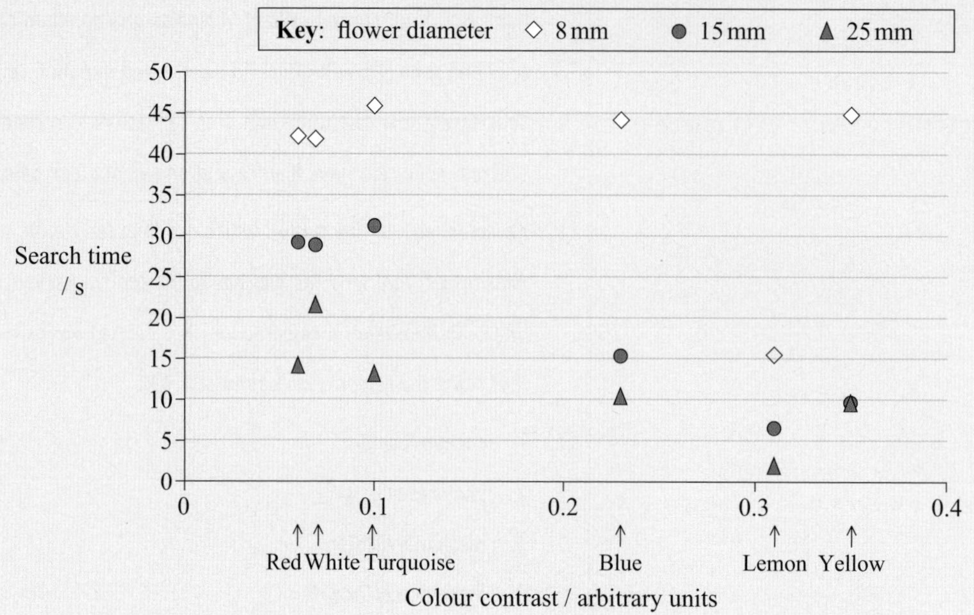

[*Source*: Spaethe, J et al. 2001. "Visual constraints in foraging bumblebees: Flower size and color affect search time and flight behavior". *Proceedings of the National Academy of Sciences*, USA. Vol 98. Pp 3898–3903. Copyright © 2001 National Academy of Sciences, USA. Reproduced with permission.]

(a) State the time it takes bumblebees to reach a blue flower of 15 mm diameter from another flower. *[1]*

(b) State the colour of flower the bumblebees find in least time. *[1]*

(c) Describe the effect of colour contrast on search time for the largest flowers (25 mm diameter). *[1]*

(d) When searching for smaller flowers, bumblebees changed the strategies used for larger flower detection. Evaluate this hypothesis using the data. *[3]*

(e) Suggest the type of behaviour shown by bumblebees in this experiment and how it can affect their chances of survival. *[1]*

[Taken from higher level paper 3, time zone 2, May 2009]

How do I approach the question?

The first two parts of the question address your ability to read the graph properly. With a more complex graph, you should use a square to find the appropriate values, but it is not necessary here.

(a) From the key for size and the axis for colour contrast, it is easy to see that 15s (always remember units!) is the time to reach blue flowers.

(b) The dark triangle for lemon flowers is clearly lower than all others.

(c) You can either name the relationship as "negative correlation" or use a sentence such as "the greater the colour contrast, the shorter the search time". Make sure that you use the terms labelled on the axis, namely colour contrast and search time.

(d) Consider the implications and limitations of the data in its support of the hypothesis. You can start by stating that the hypothesis seems to be supported as it seems to work for larger flowers, but the time is always longer for smaller flowers. You must state a possible reason, such as that this could be due to the fact that smaller flowers are more difficult to see from a distance and that another strategy must be used (for example, scent). A limitation that you could state is that lemon flowers always have a low search time, so they must provide another type of stimulus to bumblebees. Another limitation is that the flower density is unknown.

(e) There is usually more than one possible answer for a suggestion, providing that the answer is biologically possible and, in this case, that you state how it can affect the bumblebees' chances of survival. You must name the type of behaviour as innate or learned. In either case, you must mention that it will help the bumblebees to find more food.

Which areas of the syllabus is this question taken from?

- Distinction between innate and learned behaviour (E.3.1)
- Analysis of data from invertebrate behaviour in terms of chances of survival (E.3.3)

This answer achieved 2/7

1 The relationship is not described or named.
2 This does not relate to the data.

(a) From a turquoise flower to a blue flower it takes approximately 15–16 seconds. ✔

(b) Lemon flower. ✔

(c) The colour contrast of the large flowers (25 mm) makes their time to search for them quicker. 1

(d) The hypothesis is true since different behavioural strategies identify and communicate different signals → for example the waggle dance 2 can identify the direction of small/large flowers.

(e) The waggle dance 2—this affects their chances of survival since it points the bumblebees into specific directions of flowers to feed off.

This answer achieved 4/7

1 This describes the data but does not establish a link between cause and effect.
2 All possibilities are not explored.

Although not complete, this answer contains correct elements.

(a) 16 seconds ✔

(b) Lemon ✔

(c) Lemon and yellow are found quicker than blue, red, white and turquoise. Brighter yellow found first. ❶

(d) The bumblebees did not change their strategy so the hypothesis is wrong. Again, they found lemon flowers much quicker than others. ✔ They did prioritise the lemon colour in the search more, however. ❷

(e) Learned behaviour: through learning chances of survival increase as plants are easily found and therefore more food collected. ✔

This answer achieved 7/7

This answer presents enough correct elements to gain all marks.

(a) 15.5 seconds ✔

(b) Lemon ✔

(c) The general trend is that as colour contrast increases, the search time decreases ✔ (negative correlation).

(d) The search time for smallest flowers (8 mm) remained almost constant, it did not follow the same trend as the big ✔ flowers' time decreased. Therefore it could be possible that bees have different strategies: ✔ however more data and research is required to establish a causal link, for example qualitative data on the behaviour of bees. However, for the middle flowers (15 mm) the trend is the same as trend for bigger flowers, this implies that the strategy is the same. ✔

(e) This could be foraging behaviour, chances of survival are increased if a minimum amount of energy can be spent on looking for food to optimize energy intake. ✔

18. Option F: Microbes and biotechnology

Microbes and biotechnology

You should know:

- the use of reverse transcriptase
- basic techniques for gene therapy.

You should be able to:

- explain how reverse transcriptase is used in molecular biology
- distinguish between somatic and germ line therapy
- outline the use of viral vectors in gene therapy
- discuss the risks of gene therapy.

Examples

1. Outline the use of a viral vector in gene therapy. *[3]*

An example could be treatment of SCIDS (an immune deficiency condition). Some white blood cells or bone marrow cells are removed from the SCIDS patient. These cells become genetically modified when genes that express ADA (adenosine deaminase) are inserted into their chromosomes, by means of a viral vector. The treated cells are then returned to the patients where ADA production can occur.

2. Explain, with the use of a specific example, how reverse transcriptase is used in molecular biology. *[3]*

Mature mRNA (in which introns have been removed) for insulin could be used as a template to produce a complementary segment of DNA with the use of reverse transcriptase. This DNA can then be inserted in the plasmid of a bacteria that has been previously split with a specific endonuclease (restriction enzyme). When these bacteria undergo binary fission, they also copy their insulin gene containing plasmid, and the bacteria can be used to produce insulin, which can easily be extracted from their environment.

Be prepared

- It is important that you expand your answers enough to imply that reverse transcriptase uses RNA as a template to produce DNA and that RNA is not affected in the process.
- The distinction between somatic gene therapy and germ line therapy must state that desired genes are inserted into somatic (body) cells and cannot be passed on to the next generations for somatic gene therapy, whereas they are inserted in sperm cells or eggs in germ line gene therapy and can be passed on to the next generations through sexual reproduction.

Diversity of microbes (continued)

You should know:

- about the classification of living organisms into three domains
- the diversity of habitat in the Archaea
- the diversity of structure in Eubacteria
- properties of aggregates of bacteria
- cell wall characteristics of Gram-positive and Gram-negative Eubacteria
- general features of virus diversity
- general features of diversity in selected eukaryotes.

You should be able to:

- outline the classification of living organisms into Archaea, Eubacteria and Eukarya
- explain the reasons for the reclassification of living organisms into three domains
- distinguish between the organisms of the three domains using histones, introns, size of ribosomes, structure of cell walls and cell membranes
- outline the diversity of habitat of the Archaea as exemplified by methanogens, thermophiles and halophiles
- outline the diversity of Eubacteria, including shape and cell wall structure
- state that some bacteria form aggregates, such as biofilms, that show characteristics not seen in individual bacteria
- compare the structure of the cell walls of Gram-positive and Gram-negative bacteria
- outline the diversity of structure in viruses, including naked capsid versus enveloped capsid, DNA versus RNA, and single-stranded versus double-stranded DNA or RNA
- outline the diversity of microscopic eukaryotes, as illustrated by *Saccharomyces*, *Amoeba*, *Plasmodium*, *Paramecium*, *Euglena* and *Chlorella*.

Examples

1. Outline the diversity of structure in viruses. *[3]*

   Some viruses contain DNA, others RNA; some have single-stranded and others double-stranded DNA or RNA; some have a naked capsid, others an enveloped capsid.

2. Distinguish between the cell walls of Gram-positive and Gram-negative bacteria using the table below. *[2]*

Bacteria	Peptidoglycan content
Gram-positive	Thick outer wall / high peptidoglycan content
Gram-negative	Thin wall (between two layers of membrane) / low peptidoglycan content

Be prepared

- Classification of organisms using the seven levels of the hierarchy of taxa was originally designed for Eukarya, so examples of hierarchy of taxa dealing with Archaea and/or Eubacteria may not always be suitable.
- Archaea and Eubacteria have prokaryotic cells, whereas Eukarya have eukaryotic cells.
- The basic shape of bacterial cells are cocci (spherical), bacilli (rod-shaped), vibriae (comma-shaped) and spirochetes (spiral). In some species, individual cells exhibit a typical way to group themselves, for example, as clusters (staphylo-), filaments (strepto-), couples (diplo-) and so on.
- *Pseudomonas aeruginosa* and *Vibrio fischeri* are examples of aggregate-forming bacteria.
- Quorum sensing is a type of population density monitoring by a microbe when a high enough population density is reached to trigger a characteristic not seen in individuals.
- Although not required as such by the syllabus, when dealing with named examples, it could be useful to know that some microscopic eukaryotes, such as *Saccharomyces*, belong to the kingdom Fungi, whereas others, such as *Amoeba*, *Plasmodium*, *Paramecium*, *Euglena* and *Chlorella*, belong to the kingdom Protists (Protoctists). Most autotrophic Protists are referred to as algae, whereas most heterotrophic Protists are referred to as protozoa.

Microbes and the environment

Useful terminology for this section:

- **biochemical oxygen demand (BOD)**—the quantity of oxygen used for chemical oxidation processes and respiration per volume of water and unit of time, typically $mg\,l^{-1}\,day^{-1}$
- **eutrophication**—the process of aging of a body of water characterized by an increase in autotrophic organisms and nutrients.

You should know:

- the roles of microbes in ecosystems
- the nitrogen cycle
- effects and treatment of raw sewage
- principles of fuel production from biomass.

You should be able to:

- list the role of microbes in ecosystems, including producers, nitrogen fixers and decomposers
- draw and label a diagram of the nitrogen cycle
- state the roles of *Rhizobium*, *Azobacter*, *Nitrosomonas*, *Nitrobacter* and *Pseudomonas denitrificans* in the nitrogen cycle
- outline that anaerobic conditions favour denitrification and that low pH conditions and nitrifying bacteria in legumes' root nodules favour nitrification
- explain the consequences of releasing raw sewage and nitrate fertilizer into rivers
- outline the role of saprotrophic bacteria in the treatment of sewage using trickling filter beds and reed bed systems
- state that biomass can be used as raw material for the production of fuels such as methane and ethanol
- explain the principles involved in the generation of methane from biomass, including the conditions needed, organisms involved and the basic reactions that occur.

Examples

1. State the roles of *Rhizobium* and *Nitrosomonas* in the nitrogen cycle. *[1]*

   Since there is only 1 mark for this question, you must state both roles correctly to gain it. *Rhizobium* is a nitrogen fixer, whereas *Nitrosomonas* is a nitrifying bacteria.

2. State two fuels that can be produced from biomass using microbes. *[2]*

   You only need to name each one, probably for 1 mark each. Ethanol and methane are the most common fuels produced by microbes and covered by the syllabus.

Be prepared

- Nitrogen fixation, denitrification, nitrification, feeding, excretion, active transport of nitrate ions and formation of ammonia by putrefaction should be included in a drawing of the nitrogen cycle.
- The list of roles of bacteria in the nitrogen cycle should include (i) that *Rhizobium* and *Azobacter* are nitrogen fixers, converting nitrogen (N_2) into ammonia (NH_3) using ATP, (ii) that *Nitrosomonas* and *Nitrobacter* are nitrifying bacteria, *Nitrosomonas* converting NH_3 into nitrite ions (NO_2^-), in turn converted by *Nitrobacter* into nitrate ions (NO_3^-), and (iii) that *Pseudomonas denitrificans* is a denitrifying bacterium, converting NO_3^- into N_2.
- A drawing of the nitrogen cycle can also include that industry (Haber process) and lightning can contribute to nitrogen fixation.
- The presence of pathogens (chapter 12) in bathing or drinking water, eutrophication, algal blooms, deoxygenation, increase in biochemical oxygen demand (BOD) and subsequent recovery must be included when explaining the consequences of releasing raw sewage and nitrate fertilizer into rivers.

Microbes and food production

You should know:

- the use of microorganisms in food production
- the principle of food preservation
- an example of food poisoning.

You should be able to:

- explain the use of *Saccharomyces* in the production of beer, wine and bread
- outline the production of soy sauce using *Aspergillus oryzae*
- explain the use of acids and high salt or sugar concentrations in food preservation
- outline the symptoms, method of transmission and treatment of one named example of food poisoning.

Microbes and food production (continued)

Example

1. Outline the production of soy sauce. *[2]*

Your answer should only state the main steps in the production of soy sauce. The main ingredients are salt, soybeans and water. The crushed mixture is treated with *Aspergillus oryzae* (state the name fully), which breaks down the starch to glucose and eventually to lactic acid or alcohol, and breaks down protein into small peptides or amino acids. The fermentation process takes about 6 to 8 months.

Be prepared

- *Saccharomyces* is commonly called yeast.
- *Saccharomyces* and *Aspergillus* belong to a group of organisms called fungi, which are decomposers that recycle nutrients in an ecosystem.
- Beer, wine and bread are the other types of food for which you may be asked to explain the production. In all cases you should be prepared to mention the use of *Saccharomyces*, the main ingredients (barley or other cereal, grape or other fruit, and cereal, respectively) and the general process, including fermentation.

Wastewater from factories producing polyester fibres contains high concentrations of the chemical terephthalate. Removal of this compound can be achieved by certain bacteria. The graph below shows the relationship between breakdown of terephthalate and conversion into methane by these bacteria in an experimental reactor.

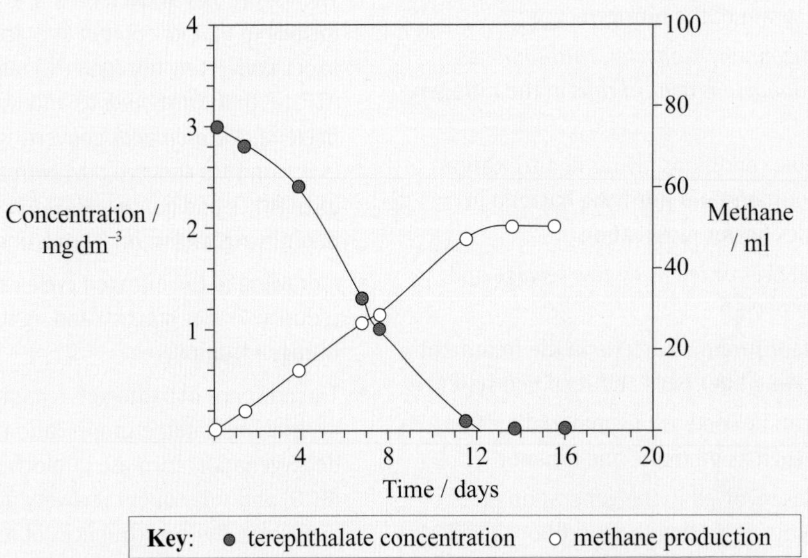

[*Source*: Wu, JH, Liu, WT, Tseng, IC and Cheng, SS. 2001. "Characterization of microbial consortia in a terephthalate-degrading anerobic granular sludge system". *Microbiology*. Vol 147. Pp 373–382. Copyright © Society for General Microbiology. Reproduced with permission.]

(a) The reactor has a volume of 12 litres. Calculate the initial amount of terephthalate in the reactor. *[1]*

(b) Describe the relationship between terephthalate concentration and methane production. *[2]*

(c) Suggest which metabolic category of bacteria could be used for the generation of methane from the degradation of terephthalate. *[1]*

(d) Evaluate the efficiency of the terephthalate breakdown into methane. *[2]*

[Taken from standard level paper 3, time zone 1, May 2009]

How do I approach the question?

Each part must be attempted, taking into account the command terms and the number of allocated marks.

(a) You have to figure out the quantity of terephthalate per unit volume to get to the answer, using your knowledge of SI (International System) units. The volume units are dm^{-3}, which are litres, and thus $3\,mg\,litre^{-1} \times 12\,litres = 36\,mg$. Always show your calculation and indicate units.

(b) A description requires that you mention more than just the inverse relationship. You could mention that both rates are approximately the same, that degradation of terephthalate and methane production are linear, or that methane production levels off from day 12, as terephthalate concentration is zero.

(c) A specific name of bacteria is not expected, but a group that accomplishes a specific function is required. Since production of methane is involved, methanogenic bacteria are most likely involved.

(d) Your answer must assess the implications and limitations, although it could contain some descriptive elements. The breakdown is efficient as all the terephthalate is processed, but it is slow as it happens over 12 days. It is also constant and the concentration used is non-toxic for the bacteria, otherwise the process would stop.

Which areas of the syllabus is this question taken from?

- Principles involved in the generation of methane from biomass (F.2.8)

This answer achieved 2/6

1 Only the concentration on the graph has been taken into account, not the volume of the reactor.

2 The description is incomplete; two elements should be mentioned in the answer.

3 There is no evaluation, only a description.

(a) The initial amount of terephthalate in the reactor is $3\,mg\,dm^{-3}$ ❶

(b) As the concentration of terephthalate decreases, the production and amount of methane is increased. ✓ ❷

(c) Decomposing bacteria can be used for the degradation of terephthalate.

(d) According to the graph, the efficiency of the terephthalate to break down into methane is eventually constant ✓ at day 14. ❸ This is the optimal point where terephthalate can be broken down.

This answer achieved 3/6

1 The figures are correct, but the concentration units have not been taken into account.
2 The breakdown rate is not evaluated.

The essence of the process has been understood, although answers could have been expanded.

(a) $0.3 \times 12 = 3.6$ litres ❶ terephthalate.

(b) As the amount of terephthalate concentration decreases, the amount of methane production increases. ✔ They are equal in amounts on the seventh day.

(c) Methanogens. ✔

(d) The efficiency of the terephthalate breakdown into methane is very efficient. ✔ The decrease/increase relationship (b) is steady, ❷ and the methane production actually stops and remains at 50 ml near the end.

This answer achieved 6/6

Complete answers have been provided, showing a clear understanding of the process.

(a) $3\,mg\,dm^{-3} \rightarrow 3\,mg/L$; $3 \times 12 = 36\,mg$ ✔
$dm^3 = L \rightarrow 12\,L$

(b) As the methane production increases, the terephthalate concentration decreases ✔ (because bacteria break down terephthalate and produce methane), until there is no more terephthalate and the concentration of methane has peaked and levelled out. ✔

(c) Bacteria called methanogens ✔ can be used for degradation of terephthalate.

(d) The breakdown of terephthalate is very efficient. ✔ In the 0 day the concentration of the chemical was very high; the breakdown starts and it is visible how the concentration of terephthalate decreases, and at the same time the conversion (production) into methane is increasing. ✔ Thus when breakdown happens methane levels are increased; this shows the efficiency.

Explain how chemicals may be used in food preservation. *[4]*

[Taken from standard level paper 3, time zone 1, May 2009]

How do I approach the question?

Your explanation should present not only the chemical, but also its mode of action and a related example. It is likely that at least two chemicals could be required for 4 marks, but since the syllabus covers acids, salt and sugar, you should use these three to construct your answer. For example, salt is a chemical used in food preservation, as it draws water out of microbes, thus killing them, as in salted meat and kippers. Mentioning osmosis is essential in your answer. The same principle applies with sugar in fruit preservatives. Weak acids, such as the acetic acid in vinegar, are also used because they lower the intracellular pH of microorganisms, thus preventing their growth, as in pickled onions.

Which areas of the syllabus is this question taken from?

- Use of acids and high salt or sugar concentration in food preservation (F.4.3)

This answer achieved 0/4

1 The mechanism of action (for example, osmosis) must be mentioned.

2 Only one type of chemical is mentioned.

Overall, this answer remains vague and does not address any biological principle.

Chemicals are used in order to keep away bacteria that may stay or incubate in the food. High concentration of salt is used for example because this environment is inappropriate for some bacteria to live and develop. 1 (Archaea bacteria (halophytes) may be an exception.) In this way chemicals create inappropriate habitats for bacteria to develop. 2

This answer achieved 1/4

1 The "other chemicals" should be named, as well as their action mechanism.

Chemicals can decrease the decomposition rate of food. For example, NaCl can preserve food by reducing water concentration ✔ in food (water helps in decomposition) and elongating the time for food to last, hence preservation. Other chemicals 1 can be used to slow decomposition also.

This answer achieved 3/4

1 More details and an example would improve this part of the answer.

Although this answer could be more complete and incorporate examples, it addresses the requirement of an explanation.

Acidic chemicals can be added to food to create a low-pH environment, ✔ in which microbes and bacteria are unable to survive. ❶ Adding large amounts of salt and sugar to foods can preserve them as well, by drawing water out ✔ of microbes/bacteria (due to osmosis, the passive diffusion of water molecules across a semi-permeable membrane), as in strawberry jam ✔ where bacteria/microbes are unable to survive if they have not been killed by the heat while preparing the jam.

19. Option G: Ecology

Community ecology

You should know:

- factors that affect the distribution of plant and animal species
- principles of sampling methods
- the meaning of the niche concept
- the main types of interactions between organisms.

You should be able to:

- outline the factors that affect the distribution of plant species, including temperature, water, light, soil pH, salinity and mineral nutrients
- explain the factors that affect the distribution of animal species, including temperature, water, breeding sites, food supply and territory
- describe one method of random sampling, based on quadrat methods, that is used to compare the population size of two plant or animal species
- outline the use of a transect to correlate the distribution of plant or animal species with an abiotic variable
- explain what is meant by the niche concept
- outline the following interactions between species—competition, herbivory, predation, parasitism and mutualism
- explain the principle of competitive exclusion
- distinguish between fundamental and realized niches.

Examples

1. Outline one example of herbivory. *[2]*

   As 2 marks are allocated for this question, you should name the herbivore and also the plant it feeds on. Remember to be as specific as you can when naming organisms.

2. Describe how biomass may be measured. *[2]*

   Your answer should indicate a few steps to measure the biomass of organisms. A destructive method would be to clean the sample of any other material, to dry it in an oven and then to measure the dry mass. If the wet (fresh) mass was measured previously, the ratio between dry mass and wet (fresh) mass could be used to calculate the biomass of similar samples in a non-destructive way.

Be prepared

- You should be able to recognize variations of kite diagrams, used to represent the abundance/density of organisms along a transect.
- You should be able to give two named examples of types of interactions between organisms.
- The dimensions that you should include in an explanation of the niche concept are an organism's spatial habitat, its feeding activities and its interactions with other species, but other dimensions may be presented for analysis in IB exams.
- The fundamental niche of a species is the potential mode of existence, given the adaptations of the species, whereas the realized niche of a species is the actual mode of existence, which results from its adaptations and competition with other species.
- The principle of competitive exclusion states that two species cannot coexist in the same ecosystem if their niches completely overlap.

Ecosystems and biomes

Required definitions for this section:

- **biomass**—the total dry mass of organic matter in organisms
- **gross production**—biomass of an organism/trophic level resulting from photosynthesis or assimilation of organic substances
- **net production**—biomass left in an organism/trophic level after organic substances have been used for cell respiration.

You should know:

- the relationship between gross production, net production, respiration and biomass
- the relationship between biomass, energy and trophic levels
- one method of biomass measurement
- the principles of succession
- the distinction between biome and biosphere
- characteristics of biomes in relation to climate components.

You should be able to:

- define biomass, gross production and net production
- calculate values for gross production (GP), net production (NP) and respiration (R) using the equation GP – R = NP
- describe one method for the measurement of biomass of different trophic levels in an ecosystem
- discuss the difficulty of classifying organisms into trophic levels
- explain the small biomass and low numbers of organisms in higher trophic levels
- construct a pyramid of energy, given appropriate information
- distinguish between primary and secondary succession, using an example of each
- outline the changes in species diversity and production during primary succession
- explain the effect of living organisms on the abiotic environment, with reference to the changes occurring during primary succession
- distinguish between biome and biosphere
- explain how rainfall and temperature affect the distribution of biomes
- outline the characteristics of biomes, such as desert, grassland, shrubland, temperate deciduous forest, tropical forest and tundra.

Examples

1. Explain the small biomass of organisms in higher trophic levels. *[2]*

   Although only 2 marks are allocated for this question, your answer should provide an account of causes or reasons. You can state that only 10–20% of energy is retained (80–90% lost) from one trophic level to the next, but you must mention that this is due to respiration and/or material that is egested or not absorbed to gain full marks.

2. Outline how two named organisms change the abiotic environment during primary succession. *[2]*

   Only a list of steps is required for your answer, but you must provide specific organism names and state their action, taking an example from a known ecosystem. For example, lichens produce acids breaking down an exposed rock. Loose particles released provide a medium for a moss, like *Polytrichum*, which continues the process further, enabling other plants to take root.

Be prepared

- The equation GP – R = NP relates gross production (GP), respiration (R) and net production (NP).
- Examples in data-based questions may sometimes refer to productivity instead of production. This should not influence the interpretation of data.
- For the purpose of pyramids of energy, the biomass or net production of a trophic level is represented by its energy value, so units for pyramids of energy are kJ m^{-2} yr^{-1}.
- Succession is the series of community species composition over time due to and causing abiotic changes.
- A primary succession takes place when no prior community is established in an area, whereas a secondary succession takes place when an environmental change has disturbed the composition of the community in an area. Explanations will involve primary successions.
- The effects of living organisms on the abiotic environment during succession should include soil development, accumulation of minerals and reduced erosion.

Ecosystems and biomes (continued)

- The word "climate" refers to annual rainfall and temperature, whereas the word "climax" refers to the final stage of a succession for a given area. Do not confuse the terms.
- You can use climatographs to illustrate the annual quantity and pattern of rainfall and temperature and their interaction on the distribution of biomes.
- When describing a biome, include annual rainfall, annual temperature and general characteristics of vegetation (which depends on the main climatic factors, annual rainfall and temperature).
- For the purpose of the syllabus, the biosphere is defined as the totality of biomes and ecosystems, although some authors also define it as planet Earth, the environment hosting life.

Impact of humans on ecosystems

Required definitions for this section:

- **biomagnification**—a process in which chemical substances become more concentrated at each trophic level.

You should know:

- how to use the Simpson diversity index
- reasons for the conservation of biodiversity
- examples of impacts and control of alien species
- the cause and consequences of biomagnification
- the general effects of chlorofluorocarbons (CFCs) on the ozone layer
- the general effects of ultraviolet (UV) radiation.

You should be able to:

- calculate the Simpson diversity index for two local communities
- analyse the biodiversity of the two local communities using the Simpson diversity index
- discuss reasons for the conservation of biodiversity, with ethical, ecological, economic and aesthetic arguments, using rainforest as an example
- list three examples of the introduction of alien species that have had significant impacts on ecosystems
- discuss the impacts of alien species on ecosystems
- explain the cause and consequences of biomagnification, using a named example
- outline the effects of ultraviolet (UV) radiation on the stratosphere, living tissues and biological productivity
- outline the effect of chlorofluorocarbons (CFCs) on the ozone layer.

Examples

1. Explain, using a **named example**, the cause and consequence of biomagnification. *[3]*

   You have to be aware of the definition of biomagnification to be able to answer this question. Establishing a list of important terminology prior to the review for exams is a useful procedure to prepare for an explanation based on a key word, such as biomagnification. Your answer must contain the specific name of the substance (for example, DDT) and its usage (for example, pesticide to control mosquitoes spreading malaria); then add the cause of biomagnification (for example, accumulates in fat tissue because it is fat-soluble). Finish by the consequence (for example, since it is not broken down by metabolism, it becomes more concentrated in each trophic level, thus causing more damage, if not death, to organisms higher in the food chain).

2. Outline the effects of ultraviolet radiation on living tissues. *[2]*

   A simple list containing at least two valid elements should enable you to gain all marks. Ultraviolet light penetrates tissues; it causes gene mutations; this leads to uncontrollable cell division; it can cause skin cancer.

Be prepared

- A formula for the Simpson diversity index is

$$D = \frac{N(N-1)}{\Sigma n(n-1)}$$

where D = diversity index, N = total number of organisms of all species found and n = number of individuals of a particular species.

Impact of humans on ecosystems (continued)

- In a list of introduced alien species that had a significant impact on an ecosystem, you should include: one named example of biological control, such as the introduction of the beetle *Galerucella* in North America to control the purple loosestrife plant (*Lythrum salicaria*); one example of accidental release, such as the starling (*Sturnus vulgaris*) in North America; and one example of deliberate release, such as the cane toad (*Bufo marinus*), an invasive species, in Australia.
- You must use the following examples in an explanation of biomagnification: mercury in fish, and organophosphorus pesticides, DDT or TBT (tributyl tin) in ecosystems.
- Be aware that ozone in the stratosphere absorbs ultraviolet (UV) radiation only to a certain limit.
- In answers to questions about the effect of UV radiation on organisms, mention that UV radiation causes damage to the most exposed cell molecules, including DNA, which can translate into an increase of skin cancer and cataracts, and a decrease of productivity by terrestrial and aquatic organisms.
- Mention that chlorofluorocarbons (CFCs) are molecules found in older refrigeration systems, propellants and industrial products, such as styrofoam, while outlining their effects on the ozone layer.

Seasonal changes of heterotrophic plankton biomass were measured in the western arctic Pacific during a one year period. The mesozooplankton, whose size is greater than 330 µm, was formed mainly by copepods. The microzooplankton, ranging from 10 to 200 µm, comprised mainly of ciliates and flagellates. Heterotrophic nanoflagellates (HNF), size range 2 to 10 µm, are organisms that feed on small flagellates and bacteria. The results are shown below.

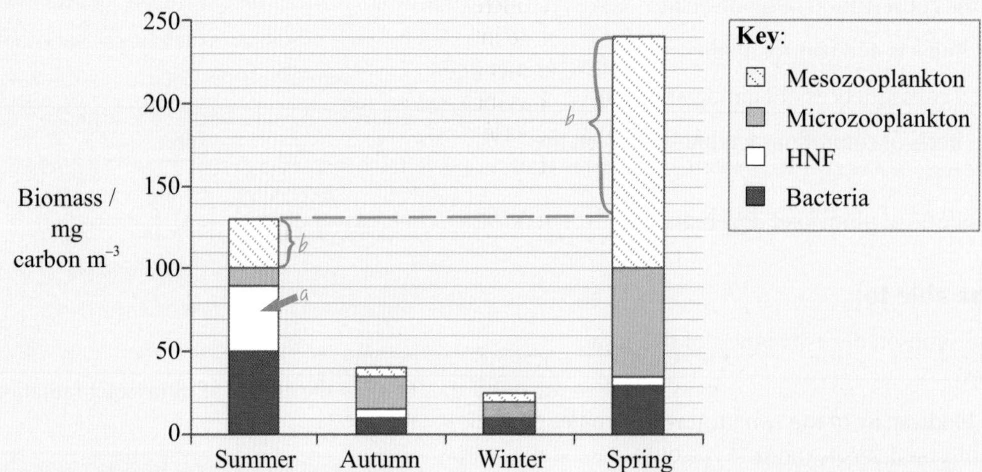

[*Source:* Shinada, A et al. 2001. "Seasonal dynamics of planktonic food chain in the Oyashio region, western subarctic Pacific". *Journal of Plankton Research*. Vol 23, number 11. Pp 1237–1248. Copyright © Oxford University Press. Reproduced with permission.]

(a) State the biomass of HNF found in this region in summer. *[1]*

(b) Calculate the percentage increase in mesozooplankton from summer to spring. Show your working. *[2]*

(c) Suggest how the seasonal changes cause the differences in biomass of heterotrophic plankton. *[3]*

[Taken from higher level paper 3, time zone 2, May 2009]

How do I approach the question?

Your first step in approaching this question, as with all other data-based questions, is to read the introductory paragraph (stem) and highlight the relevant information. In this case, it could be useful to transcribe the sizes of organism categories by the key. The size of bacteria is not provided since it should be known from the syllabus (see chapter 7). It then appears that organisms are ranked from the largest to the smallest, top to bottom. Some words or terms may be new or unknown (for example, ciliates and flagellates): highlight them with another colour. Do not worry if you do not know what they mean, as you may only have to refer to them; other times the relevant information about them (for example, size category) will be provided.

(a) What and where? In this type of graph, a complete column represents all types of heterotrophic plankton and HNF is represented in the white areas only; the summer is the first bar on the left. The height of the area indicated by arrow "a" has to be measured. A square should normally be used to transfer the limits of the area on the *y*-axis, but in this case both ends of the bar correspond to the graph's horizontal lines. Check the value of the increments on the *y*-axis: here they are 10 mg carbon m^{-3}. The answer is therefore 40 mg carbon m^{-3}. Do not forget to write the units.

(b) Many data-based questions ask for a percentage increase and all students should practise this calculation. Mesozooplankton is represented by the top box in each bar. Measure the values for summer and spring accurately (pick the correct columns, use a square and check grid increments). Since the question says "from summer to spring", the absolute increase is the value for spring (the last one) minus the value for summer (the first one). (For some questions, the answer could be negative, thus indicating an actual decrease.) The percentage increase is the absolute increase divided by the value for summer, times 100. As a formula, it would be:

$$\% \text{ increase} = \frac{\text{spring} - \text{summer}}{\text{summer}} \times 100 = \frac{140 - 30}{30} \times 100$$
$$= 367\%$$

Again, do not forget the units: "%".

(c) The first step is to limit the scope of the question, perhaps highlighting the important words: seasonal changes, and heterotrophic plankton (the complete bars on the graph). Knowledge of trophic relations could be an element of the answer. Heterotrophs are organisms that obtain organic molecules from other organisms (see "Useful terminology for this section" on page 57). Therefore, heterotrophic plankton, due to the size of its components, represents low-level consumers in a food web. Plankton abundance depends on the abundance of producers, but also on higher-level consumers and seasonal variations. The command term for this question is "suggest" and there are 3 marks. Therefore examiners are looking for three possible answers, providing that they are supported by some biological background. The answer could be organized by bulleted items, each one mentioning a factor and how it would act (for example, increased light, so more biomass of producers is available). Also the answer should contain more than three items, in case some do not score marks.

Which areas of the syllabus is this question taken from?

- Factors that affect the distribution of organisms (G.1.2)
- Interactions between species (G.1.6)
- Gross production, net production and biomass (G.2.1)
- This question also draws on knowledge from the core, fundamental biological principles, required mathematical skills (chapter 6) and aim 4 of group 4 subjects (develop an ability to analyse, evaluate and synthesize scientific information).

This answer achieved 2/6

1 This is a suggestion that would explain a larger biomass.

1 The student has used the total value for mesozooplankton in the spring instead of only the increase.
2 The rationale for this calculation is hard to follow. If x is the value for summer, it is difficult to understand why it is multiplied by the spring value.
3 The question is about heterotrophic plankton, not only bacteria.
4 This would apply to terrestrial conditions, but not to arctic waters.
5 This is a description. It does not suggest how.

(a) 40 mg carbon m^{-3} ✓

(b) 140 (1) → 100%

30 → x

140x = 100 × 30

$140x = \frac{3000}{140}$ (2)

x = 21.4%

Answer: 21.4%

(c) Bacteria (3) have a larger biomass in summer and spring as a result of hot moist conditions. (4) HNF does not appear in winter according to the graph and show little indication of biomass in spring and autumn, but a biomass of 40 mg carbon m^{-3} in summer. (5) This is probably because there is more bacteria in summer to feed on ✓ (1) and less microzooplankton to feed on it.

This answer achieved 2/6

1 This expression makes sense if we read "new value" as "increase".
2 This is a valid suggestion for how their population would increase.

1 The student used only the value of the upper limit for HNF. The value of the lower limit should have been subtracted: 90 – 50 mg carbon m^{-3} = 40 mg carbon m^{-3}.
2 Again, values of the upper limit are used, this time for mesozooplankton. This demonstrates an inability to read this graph properly, leading to a wrong answer.
3 There is no need to keep two decimal places, as the values show three significant figures.
4 There is no suggestion for the reason for their increase (for example, light, photosynthesis, higher temperature).

(a) 90 (1) mg carbon m^{-3}

(b) $\frac{\text{new value} - \text{old value}}{\text{old value}}$ (1) × 100

$= \frac{240 - 130}{130} \times 100$ (2)

= 184.62% increase (3)

(c) In summer and spring, the biomasses of the flagellates and bacteria, which HNF feed on, increase. (4) Since their food source has increased, the biomass of HNF increases, ✓ due to increased resources. In winter and autumn months the temperature is colder, so less flagellates and bacteria survive. As a result, there are fewer HNF, as they have less food, so fewer survive, and their biomass decreases. ✓ (2)

This answer achieved 6/6

1. Values for each season are found, the correct formula is applied, leading to a correct answer.
2. Specifying the number of significant figures is not necessary, but it shows an effort to write a complete answer.
3. This opening statement relates the question to the understanding of a food chain/web.

1. Strictly, "× 100" should have been written on both sides of the equation.

This is a clear answer that shows a good ability to read the graph and the application of knowledge to a new situation.

(a) 40 mg carbon m^{-3} ✓

(b) 30 = summer; 140 = spring

increase of 110

$\frac{110}{30} = 3\frac{2}{3} \times 100$ ✓ ❶

= 367% increase ❶ (3 s.f.) ❷

367% (3 s.f.) increase in mesozooplankton from summer to spring. ✓

(c) The majority of the heterotrophic plankton will feed on phytoplankton. ❸ These are photosynthetic organisms which undergo a seasonal bloom in spring and summer, ✓ vastly increasing their numbers, as during the winter, the arctic Pacific is dark all day with no sunlight, ✓ so few phytoplankton can survive. This means the consumers which eat phytoplankton follow the same population increases and decreases, with the availability of food. ✓

(a) Define the terms (i) gross production and (ii) net production. *[2]*

(b) Using the table below, compare the characteristics of the following biomes. *[3]*

Biome	Temperature	Moisture	Characteristic of vegetation
Desert			
Tropical rainforest			
Tundra			

[Taken from standard level paper 3, time zone 2, May 2009]

How do I approach the question?

This question is totally based on recall of factual knowledge, so there is no other choice than writing what can be remembered!

(a) Probably 1 mark for each definition is allocated. You should mention important words like biomass, photosynthesis and cell respiration. Writing a previously learned definition (see those in the text) is probably a better strategy than creating a new one during the exam!

(b) Since 3 marks are allocated for this question, you can assume that three elements in the same category are required for each mark. (In this case, one mark was allocated for a full correct line or a full correct column.)It would be preferable, although the spaces are not very large, to specify ranges of values or to outline details instead of writing only one or a few words in each space (for example, "desert: average rainfall less than 25 cm year^{-1}"). Ideally, comparison words should be used for each category, although the presentation in a table is by itself a form of comparison.

Which areas of the syllabus is this question taken from?

- Define gross production, net production and biomass (G.2.1)
- Outline the characteristics of six major biomes (G.2.11)

This answer achieved 1/5

1 Enough details are provided for all three characteristics only for this biome.

1 These definitions do not relate to biomass/organic matter or to respiration, neither do they relate to the formula GP – R = NP.

2 This does not relate to rainfall and no details are provided.

3 This does not relate to low precipitation. Melting snow is only for a very limited time and is not an abundant source of water.

4 Bogs are correct, but this generally does not relate to the type of vegetation/absence of trees (trees can have flowers too).

(a) (i) Gross production: The total production

(ii) Net production: The mean production (1)

(b)

Biome	Temperature	Moisture	Characteristic of vegetation
Desert	High temperatures Low temperatures in cold deserts	Extremely low (2)	Presence of cacti plants, scarce vegetation
Tropical rainforest (1)	High temperatures, not much difference from day to night	Extremely high due to heavy rainfall	Lush vegetation, various evergreen trees and plants ✓
Tundra	Low all year around except for a few weeks in summer	High in summer due to melting of snow (3)	Presence of anaerobic bogs, lots of flowers in summer (4)

This answer achieved 2/5

1 Relates to the formula GP – R = NP.
2 Accurate description of vegetation characteristics for all three biomes.

1 Should relate to biomass/organic matter instead. Photosynthesis is not mentioned either.
2 "Biomass" would be preferable.
3 Only values are mentioned, without an appreciation of fluctuations for all three biomes.
4 Correct characteristics are mentioned, but the quoted values are too far from an acceptable range and are contradictory.

(a) (i) Gross production: Amount of energy 1 produced in total by an organism.

(ii) Net production: Left over 2 of what is used from gross production after respiration. ✓ 1

(b)

Biome	Temperature	Moisture	Characteristic of vegetation 2
Desert	Summer 50°C winter −30°C	Very dry less than 300 ml of rain per year	Plants can retain water, don't have leaves (less water loss). Adapted to extreme temperatures ✓
Tropical rainforest	20–25°C	A lot of rain per year 500–1500 ml	Dark green leaves, use up sunlight best; very tall trees; water is not hard to obtain; not complicated root systems or bark for water retaining ✓
Tundra	−30°C up to 30°C 3	Dry and cold 300–500 ml of rain per year 4	Not very many to any trees (too little water) Bushes and short vegetation because of permafrost, not many plants can grow ✓

This answer achieved 4/5

1. This is a valid definition because "organic matter" and "plants" are mentioned.
2. Each line in this table contains just enough details to outline properly the characteristics of each biome.

1. This should relate to biomass/organic matter instead. Photosynthesis is not mentioned either.
2. Nothing would remain if GP was subtracted. Respiration removes organic matter, it does not produce it. This contradicts the concept of "production" and the answer cannot be considered as a valid definition.

(a) (i) Gross production: Total of organic matter produced by plants in an ecosystem. 1

(ii) Net production: Total of gross production remaining after subtracting it from the organic matter produced by respiration. 1 2

(b)

Biome 2	Temperature	Moisture	Characteristic of vegetation
Desert	Hot during the day. Very cold nights	Very dry	Few vegetation. Quickly grow after rainfall. Store water. Cacti ✓
Tropical rainforest	Hot humid temperatures	Wet moist atmosphere	Huge diversity of species. Long tall trees, shrubs, evergreens. Huge leaves ✓
Tundra	Very cold	Some precipitation, mainly as snow	Mosses, lichens, small short vegetation to withstand winds and small to keep the heat ✓

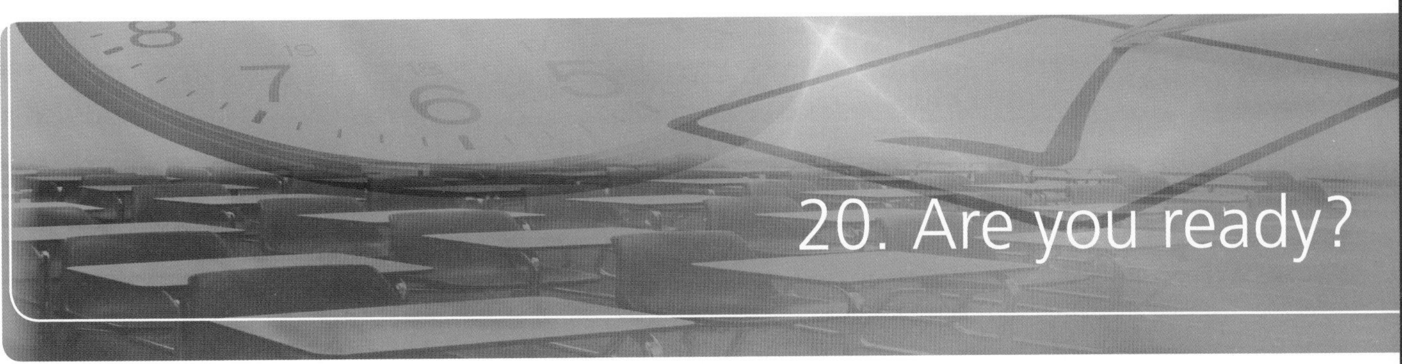

20. Are you ready?

In this chapter, we have added all of the paper 2 and most of the paper 3 exam questions from May 2009 and November 2009 that were not used in the earlier chapters. This will provide you with the opportunity to practise some of the skills that you have picked up as you have worked through this book. **Note that not all parts of the subject syllabus have been covered by these questions.** In some cases, questions are similar and demonstrate the ways in which exam authors can vary questions based on the same assessment statements. Owing to space limitations, examples of multiple-choice questions as well as some paper 3 data-based questions have not been included. Otherwise, the questions have been organized by topic or option according to the chapters of this book.

In the following tables, "Reference" indicates paper number (P2 for chapters 6–12, P3 for options chapters 13–19), level (HL or SL), time zone (TZ1 = Americas (English), TZ2 = other regions and French, Spanish and German worldwide), session (May 2009 or November 2009), and question number and/or part of question, respectively. The information paragraph and data of a question are referred to as the "stem".

Chapter 6: Statistical analysis, mathematical requirements and data-based questions

Subtopic	Marks	Reference
Type II diabetes is having an impact on the health of many individuals worldwide. The condition is characterized by elevated levels of both insulin and glucose in the bloodstream. Some animals produce an insulin-degrading enzyme (IDE) which breaks down the insulin molecule. In an attempt to develop a model of type II diabetes, genetically modified mice have been developed. In these mice, both copies of the IDE gene have been removed (IDE–/–) and the enzyme is not produced. The bar chart below shows the mean concentration of insulin in the bloodstream of IDE–/– mice and that of control mice (IDE+/+).		P2 HL N09 q1stem

Blood insulin/ng ml^{-1}

2.0
1.5
1.0
0.5
0.0

IDE +/+ IDE –/–

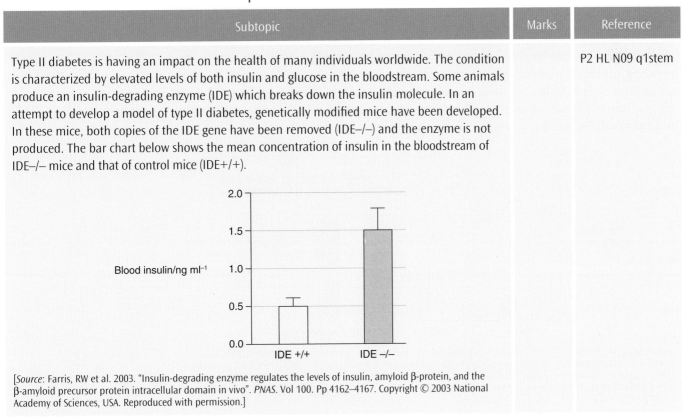

[*Source*: Farris, RW et al. 2003. "Insulin-degrading enzyme regulates the levels of insulin, amyloid β-protein, and the β-amyloid precursor protein intracellular domain in vivo". *PNAS*. Vol 100. Pp 4162–4167. Copyright © 2003 National Academy of Sciences, USA. Reproduced with permission.]

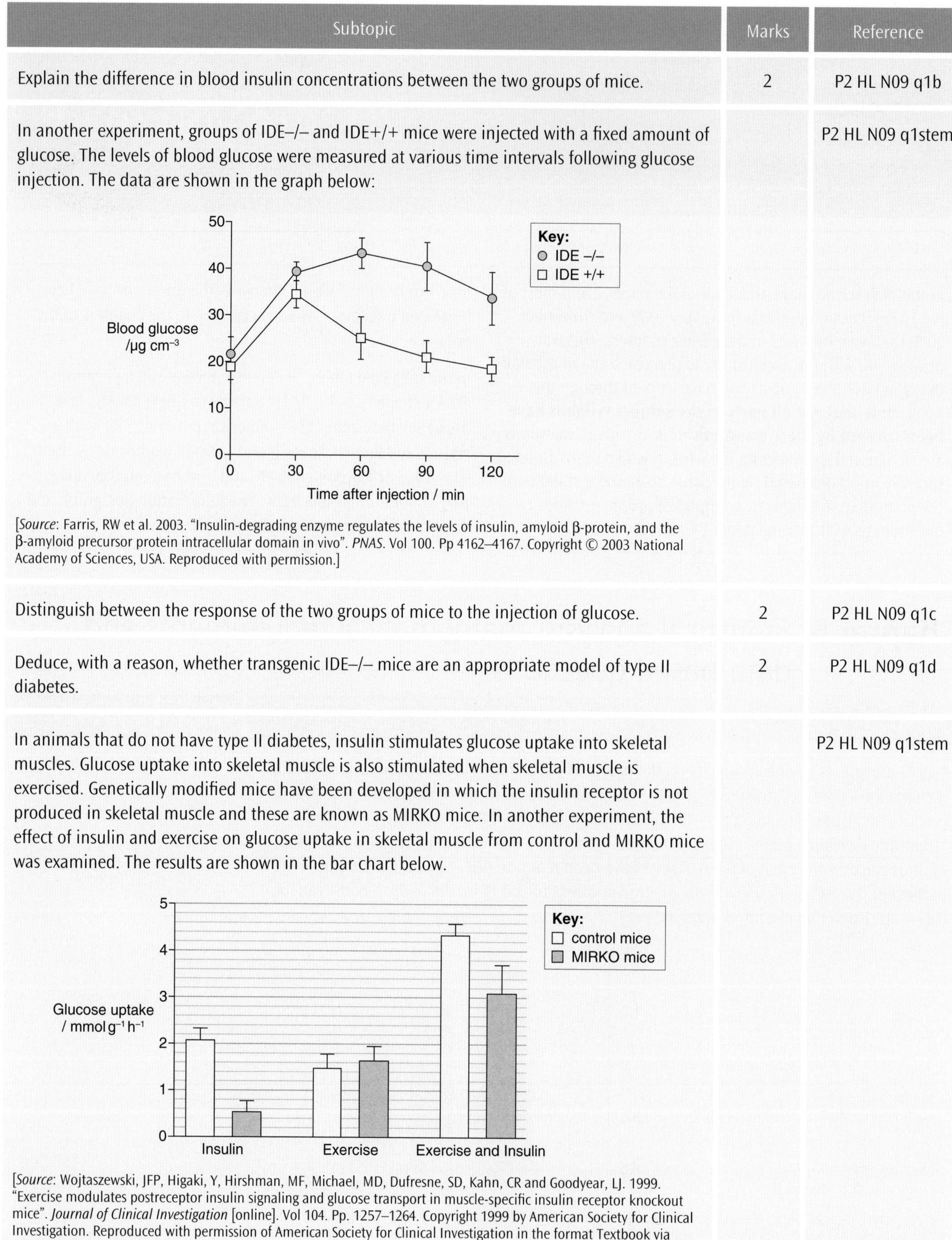

Subtopic	Marks	Reference
Explain the difference in blood insulin concentrations between the two groups of mice.	2	P2 HL N09 q1b
In another experiment, groups of IDE–/– and IDE+/+ mice were injected with a fixed amount of glucose. The levels of blood glucose were measured at various time intervals following glucose injection. The data are shown in the graph below: [*Source*: Farris, RW et al. 2003. "Insulin-degrading enzyme regulates the levels of insulin, amyloid β-protein, and the β-amyloid precursor protein intracellular domain in vivo". *PNAS*. Vol 100. Pp 4162–4167. Copyright © 2003 National Academy of Sciences, USA. Reproduced with permission.]		P2 HL N09 q1stem
Distinguish between the response of the two groups of mice to the injection of glucose.	2	P2 HL N09 q1c
Deduce, with a reason, whether transgenic IDE–/– mice are an appropriate model of type II diabetes.	2	P2 HL N09 q1d
In animals that do not have type II diabetes, insulin stimulates glucose uptake into skeletal muscles. Glucose uptake into skeletal muscle is also stimulated when skeletal muscle is exercised. Genetically modified mice have been developed in which the insulin receptor is not produced in skeletal muscle and these are known as MIRKO mice. In another experiment, the effect of insulin and exercise on glucose uptake in skeletal muscle from control and MIRKO mice was examined. The results are shown in the bar chart below. [*Source*: Wojtaszewski, JFP, Higaki, Y, Hirshman, MF, Michael, MD, Dufresne, SD, Kahn, CR and Goodyear, LJ. 1999. "Exercise modulates postreceptor insulin signaling and glucose transport in muscle-specific insulin receptor knockout mice". *Journal of Clinical Investigation* [online]. Vol 104. Pp. 1257–1264. Copyright 1999 by American Society for Clinical Investigation. Reproduced with permission of American Society for Clinical Investigation in the format Textbook via Copyright Clearance Center.]		P2 HL N09 q1stem

<table>
<tr><th>Subtopic</th><th>Marks</th><th>Reference</th></tr>
<tr><td>Explain the reason for the difference in insulin-stimulated glucose uptake between control mice and MIRKO mice.</td><td>2</td><td>P2 HL N09 q1e</td></tr>
<tr><td>Distinguish between the effects of insulin alone and exercise alone on glucose uptake in skeletal muscle of MIRKO mice.</td><td>1</td><td>P2 HL N09 q1f</td></tr>
<tr><td>Evaluate, using the data, whether exercise would be an appropriate therapy for human patients with type II diabetes.</td><td>3</td><td>P2 HL N09 q1g</td></tr>
<tr><td>Limpets are small animals that feed on the green algae which grow on rocks on seashores. Oystercatchers (Haematopus bachmani) are birds that feed on limpets.

Limpet
Oystercatcher

In a study on the north-west coast of the USA, where three limpet species are common (Lottia digitalis, Lottia pelta and Lottia strigatella), the limpets were protected from the oystercatchers by large wire cages. After two years the number of limpets in this area was compared with the number of limpets in an area without cages, where oystercatchers were present.

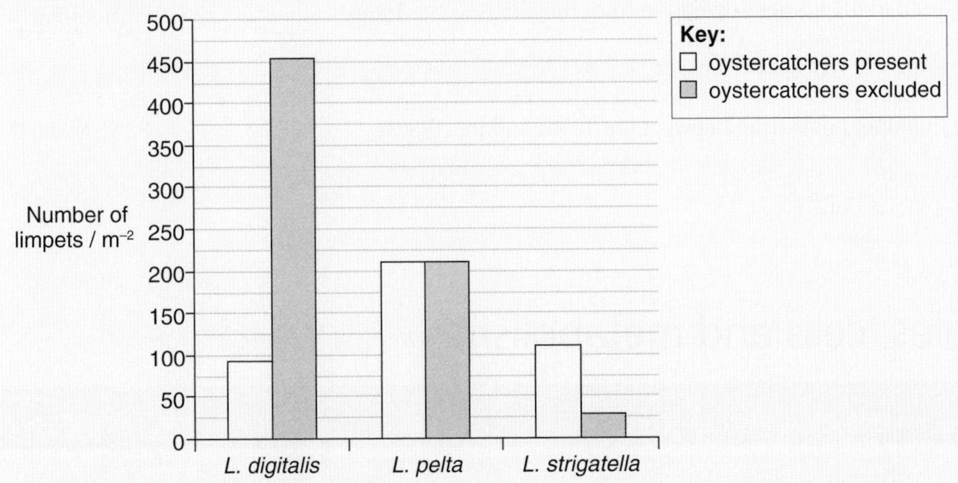

[Source: Wootton, JT. 1992. "Indirect effects, prey susceptibility, and habitat selection: Impacts of birds on limpets and algae". Ecology. Vol 73. Pp 981–991. Copyright 1992 by Ecological Society of America. Reproduced with permission of Ecological Society of America in the format Textbook via Copyright Clearance Center.]</td><td></td><td>P2 SL N09 q1stem</td></tr>
<tr><td>State the effect that the exclusion of the oystercatchers had on the total number of limpets per m^2.</td><td>1</td><td>P2 SL N09 q1a</td></tr>
<tr><td>Suggest reasons for the difference in numbers of L. strigatella between the areas where oystercatchers were present and where oystercatchers were excluded.</td><td>3</td><td>P2 SL N09 q1c</td></tr>
</table>

Subtopic	Marks	Reference
There is evidence to show that both air and water temperatures have increased over a period of time. An investigation was undertaken to determine the effect this change in climate had on the populations of another species of limpet, *Patella depressa*, around south-west England. The population of the limpet was recorded in many locations and around 30 years later, this study was repeated. The chart below compares the population in each of the locations. 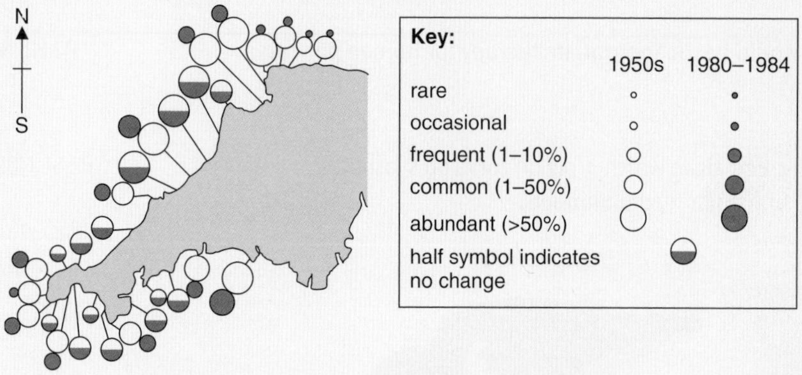 N S **Key:** 1950s / 1980–1984 rare occasional frequent (1–10%) common (1–50%) abundant (>50%) half symbol indicates no change [*Source*: Kendall, MA et al. 2004. "Predicting the effects of marine climate change on the invertebrate prey of the birds of rocky shores". *IBIS*. Vol 146. Fig 2. Pp 40–47. Copyright © John Wiley and Sons. Reproduced with permission.]		P2 SL N09 q1stem
On the map above label **one** location, (i) with the letter X, where the limpet population was abundant in the 1950s and occasional in the 1980s (1980–1984). (ii) with the letter Y, where the limpet population was abundant in both the 1950s and the 1980s (1980–1984).	1 1	P2 SL N09 q1d
Outline, using the data, the overall trend in the limpet population from the 1950s to the 1980s (1980–1984).	2	P2 SL N09 q1e
Suggest **two** reasons for the change in limpet population between the 1950s and the 1980s (1980–1984).	2	P2 SL N09 q1f

Chapter 7: Molecules, cells and metabolism

Subtopic/Question	Marks	Reference
Cell theory		
State the property of stem cells that makes them useful in medical treatment.	1	P2 SL TZ2 M09 q2a
Explain how multicellular organisms develop specialized tissues.	2	P2 SL TZ2 M09 q2b
Prokaryotic and eukaryotic cells		
Compare the relative sizes of viruses and bacteria to this cell.	2	P2 HL TZ1 M09 q3b

Subtopic/Question	Marks	Reference
Compare prokaryotic and eukaryotic cells.	3	P2 SL TZ1 M09 q2a
Draw a labelled diagram showing the ultrastructure of a typical prokaryote.	4	P2 SL TZ2 M09 q7a
Draw a labelled diagram to show the ultrastructure of *Escherichia coli*.	5	P2 HL N09 q4a
The electron micrograph below shows an *E. coli* cell. 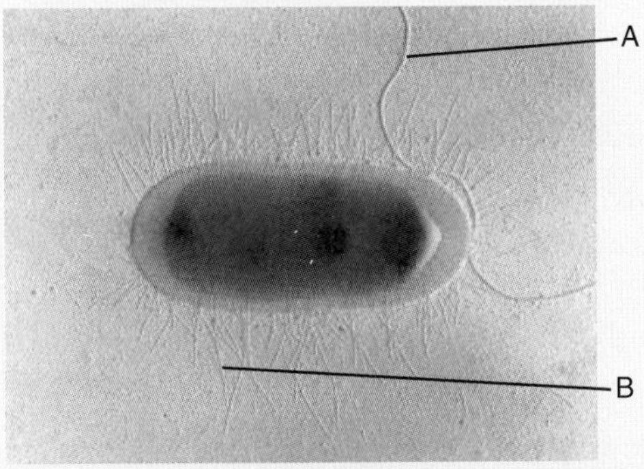 A B [Copyright © David J. Silverman University of Maryland School of Medicine. Reproduced with permission.] Identify the structures labelled A and B in the electron micrograph above and state one function of each. A: Name .... Function .... B: Name .... Function	2	P2 SL TZ1 M09 q2a
Membranes		
Draw a labelled diagram to show the structure of a membrane.	5	P2 SL TZ1 M09 q6a
Outline how vesicles are used to transport materials secreted by a cell.	6	P2 SL TZ1 M09 q6b
Explain how materials are moved across membranes of cells by active transport.	2	P2 SL TZ2 M09 q3a
Distinguish between active and passive movements of materials across plasma membranes, using **named** examples.	4	P2 HL/SL N09 q4b
Chemistry of life		
Outline the thermal, cohesive and solvent properties of water.	5	P2 HL N09 q7a
Explain how the properties of water are significant to living organisms.	9	P2 SL N09 q4c
Cell respiration		
Outline anaerobic cell respiration in plant cells.	5	P2 SL N09 q6b
Photosynthesis		
Explain the effect of light intensity and temperature on the rate of photosynthesis.	8	P2 HL TZ2 M09 q4c
Outline how **three** different environmental conditions can affect the rate of photosynthesis in plants.	6	P2 SL TZ2 M09 q7b

Chapter 8: Nucleic acids

Subtopic/Question	Marks	Reference
DNA structure		
State the type of bonds that (i) connect base pairs in a DNA molecule. (ii) link DNA nucleotides into a single strand.	 1 1	P2 SL N09 q2a
Transcription/translation		
Distinguish between DNA and RNA nucleotides by giving **two** differences in the chemical structure of the molecules.	2	P2 SL TZ1 M09 q2b
Explain the role of transfer RNA (tRNA) in the process of translation.	2	P2 HL N09 q3d P2 SL N09 q2c
Chromosomes, genes, alleles and mutations		
Outline a possible cause of Down syndrome.	4	P2 HL TZ1 M09 q5a
Describe karyotyping and **one** application of its use.	4	P2 SL N09 q7a
Genetic engineering		
Outline some of the outcomes of the sequencing of the human genome.	3	P2 SL TZ2 M09 q2a
Cloning and gene transfer & modification		
Outline a basic technique for gene transfer.	6	P2 SL TZ1 M09 q7b
Describe a technique used for gene transfer.	5	P2 SL N09 q7b

Chapter 9: Genetics

All the available questions have been used in the previous chapters.

Chapter 10: Classification, ecology and evolution

Subtopic/Question	Marks	Reference
Communities and ecosystems		
Construct a food web to show the feeding relationship between the three species of limpets, the oystercatchers and the green algae. (This question refers to P2 SL N09 q1 stem shown on page 157.)	2	P2 SL N09 q1b
State **one** role of bacteria in a soil ecosystem.	1	P2 SL TZ1 M09 q1f

Subtopic/Question	Marks	Reference
Populations		
Draw a labelled sigmoid population growth curve.	4	P2 HL N09 q6a
Evolution		
Describe how natural selection leads to evolution.	6	P2 SL TZ2 M09 q6b
Classification		
Maize belongs to the group of plants known as angiospermophyta. Distinguish between angiospermophytes and bryophytes.	2	P2 HL TZ1 M09 q2b

Chapter 11: Human physiology, part 1

Subtopic/Question	Marks	Reference
Digestion		
Researchers extracted an enzyme from the human digestive system and tested its activity at different pH values on proteins extracted from the blood of cows. The results are shown in the graph below. 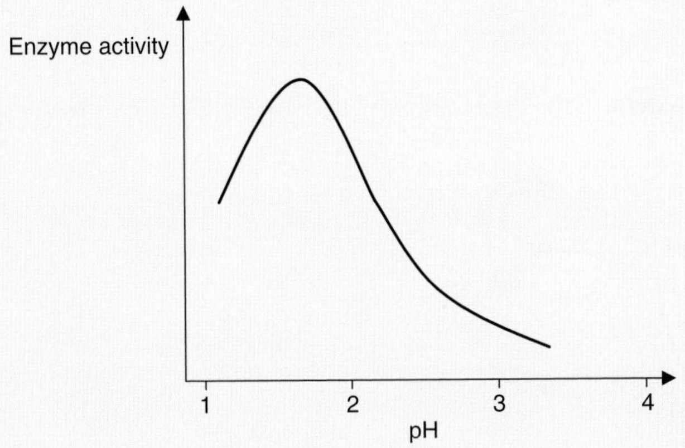 Enzyme activity 1 2 3 4 pH Deduce from where in the human digestive system this enzyme was extracted.	1	P2 SL N09 q3a
State **one** function of the large intestine.	1	P2 SL N09 q3c
Explain how the structure of the villus is adapted for absorption.	3	P2 SL N09 q3d
Gas exchange		
Draw a labelled diagram to show the human ventilation system.	4	P2 SL N09 q6a

Chapter 12: Human physiology, part 2

Subtopic/Question	Marks	Reference
Hormones and homeostasis		
Explain the concept of homeostasis, using the control of blood sugar as an example.	9	P2 SL N09 q6c
Distinguish between type I and type II diabetes.	3	P2 SL TZ1 M09 q5
State which cells secrete insulin and the organ in which they are located.	2	P2 HL N09 q1h
State the name of **one** hormone other than insulin involved in the regulation of blood glucose.	1	P2 HL N09 q1i
Defence against infectious disease		
Discuss how the HIV virus is transmitted.	2	P2 HL N09 q2c
Explain why antibiotics are ineffective against viruses.	2	P2 HL N09 q2d
Reproduction		
Draw a labelled diagram of the adult female reproductive system.	4	P2 HL/SL TZ2 M09 q5a
The diagram below shows the female reproductive system. 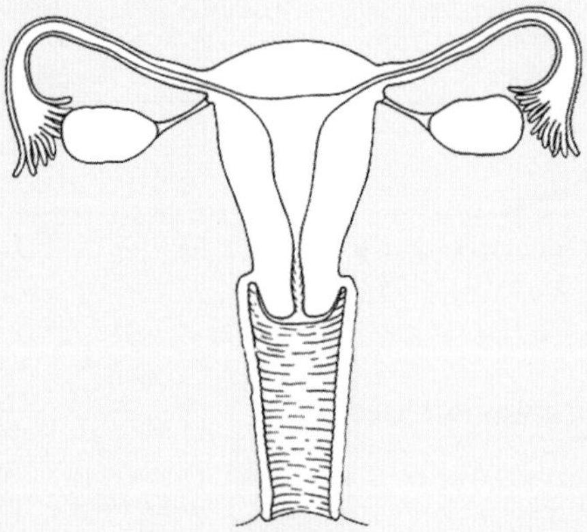 Label the diagram above with the letter U to show the uterus.	1	P2 SL N09 q4a
Outline the roles of progesterone and estrogen in the human menstrual cycle.	6	P2 HL TZ2 M09 q5b
Outline the role of luteinizing hormome (LH) **after** ovulation.	1	P2 HL N09 q4b

Chapter 13: Option A—Human nutrition and health

Question	Marks	Reference
Outline the role of the brain in how appetite is controlled.	2	P3 SL TZ1 M09 q2a
State **one** consequence of protein deficiency malnutrition.	1	P3 N09 q2a
Outline the reasons for increasing rates of clinical obesity in some countries.	3	P3 N09 q2b
Outline a method used to determine the recommended daily intake of vitamin C.	3	P3 SL TZ1 M09 q3a
Discuss the amount of vitamin C that adults should consume per day.	4	P3 SL TZ1 M09 q3b
State **one** cause of type II diabetes.	1	P3 TZ2 M09 q3ai
Outline the variation in the structure of fatty acids.	3	P3 N09 q3

Chapter 14: Option B—Physiology of exercise

Question	Marks	Reference
Draw a labelled diagram to show the structure of a skeletal muscle sarcomere.	3	P3 N09 q2a
Define *fitness*.	1	P3 SL TZ1 M09 q3a

Chapter 15: Option C—Cells and energy

Question	Marks	Reference
Outline the metabolism of glucose during glycolysis.	2	P3 TZ2 M09 q1d
Describe the effect of salt concentration on the activity of the light-dependent reactions overall.	1	P3 N09 q1a
Label the structures in the electron micrograph showing part of a cell containing a mitochondrion. 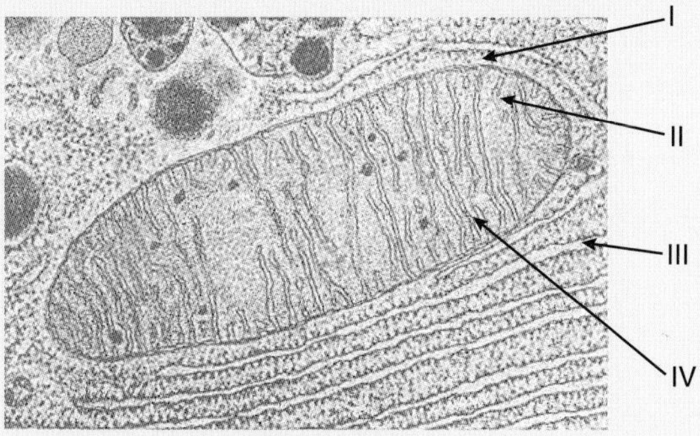 I, II, III, IV [Copyright © Tribe, M and Whittaker, P. 1972. *Chloroplasts and Mitochondria*. Edward Arnolds Publishers.]	2	P3 SL TZ1 M09 q2a

Question	Marks	Reference
Outline the relationship between the structure of the mitochondrian and its function.	2	P3 TZ2 M09 q2b
State **one** example of a fibrous protein.	1	P3 N09 q2a
Distinguish between the secondary structure and tertiary structure of proteins.	3	P3 N09 q2b
Explain the control of metabolic pathways by end-product inhibition.	4	P3 SL TZ1 M09 q3c

Chapter 16: Option D—Evolution

Question	Marks	Reference
State which theory is supported by the presence of DNA in mitochondria.	1	P3 SL TZ2 M09 q1d
Describe the processes needed for the spontaneous origin of life on Earth.	3	P3 SL TZ1 M09 q2a
Discuss the endosymbiotic theory for the origin of eukaryotes.	3	P3 SL TZ1 M09 q2b
State **one** process needed for the spontaneous origin of life on Earth.	1	P3 HL TZ1 M09 q2a
Outline the contribution of prokaryotes to the creation of an oxygen-rich atmosphere.	2	P3 HL TZ1 M09 q2b
State **two** properties of RNA that would allow it to play a role in the origin of life.	2	P3 SL TZ2 M09 q2a
State **one** difference between cultural evolution and genetic evolution.	1	P3 HL TZ2 M09 q2a
State the conclusion drawn from the Miller-Urey experiment.	1	P3 SL N09 q2a
Outline allopatric and sympatric speciation.	4	P3 HL TZ1 M09 q3a
Define the term *gene pool*.	1	P3 SL TZ2 M09 q3a

Chapter 17: Option E—Neurobiology and behaviour

Question	Marks	Reference
Define *reflex*.	1	P3 HL TZ1 M09 q2bi
Label this diagram of the ear. 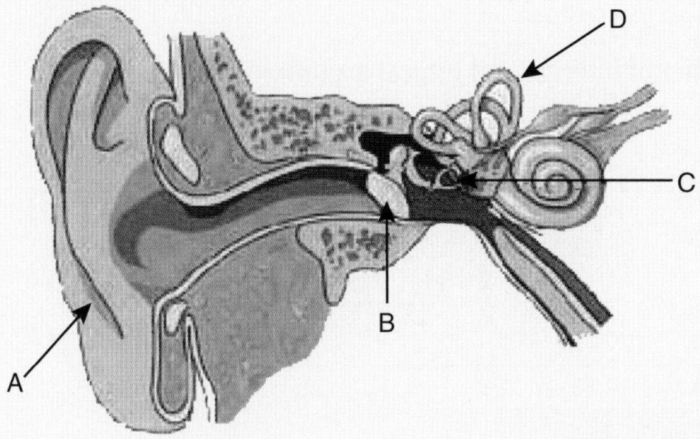 A B C D	4	P3 HL TZ1 M09 q2a
Label the **four** parts of the ear indicated on the drawing below. 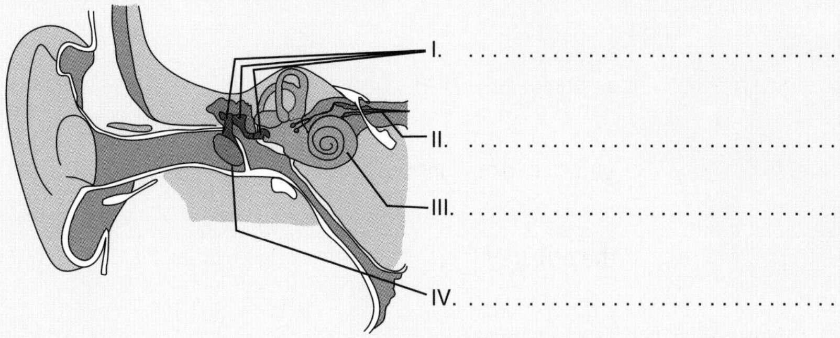 I. .... II. .... III. .... IV.	2	P3 HL/SL TZ2 M09 q2a
Outline Pavlov's experiment into conditioning of dogs.	2	P3 SL TZ2 M09 q2c
Distinguish between a conditioned response and an unconditioned response.	2	P3 SL N09 q3a
Describe an experiment investigating innate behaviour in invertebrates.	4	P3 HL TZ1 M09 q3a
State **one** effect of tetrahydrocannabinol (THC) on brain function.	1	P3 HL N09 q2c
Explain the effect of tetrahydrocannabinol (THC) on brain function.	2	P3 SL N09 q3b
Discuss the role of genetic predisposition and dopamine secretion in addiction.	4	P3 SL TZ1 M09 q3b

Chapter 18: Option F—Microbes and biotechnology

Question	Marks	Reference
Probiotics reside in intestines. State the name of **one** group of Archaea that can be found in animal intestines.	1	P3 SL N09 q1e
Discuss the risks involved in gene therapy.	2	P3 SL TZ1 M09 q2b
Discuss the risks of gene therapy including safety, conflict of interest and ethical arguments.	6	P3 HL N09 q3
State **one** example of a characteristic shown by aggregates of a **named** bacterium not present in an individual of that same species.	2	P3 SL TZ2 M09 q3a
The electron micrograph below shows a pathogen. [*Source*: Copyright © Professor Frederick A Murphy. Reproduced with permission.]		P3 SL N09 q2
Identify the type of pathogen shown in the electron micrograph, giving reasons for your answer.	2	P3 SL N09 q2a
State **one** way nucleic acids can vary in viruses.	1	P3 HL TZ2 M09 q2a
Distinguish between *Saccharomyces* and *Chlorella* in terms of mode of nutrition.	1	P3 SL N09 q3c
State **two** fuels that can be produced from biomass using microbes.	2	P3 SL N09 q3a
Explain the use of *Saccharomyces* in the production of beer.	3	P3 SL TZ2 M09 q3b
Explain the significance of *Saccharomyces* in the production of bread.	3	P3 SL N09 q3b

Chapter 19: Option G—Ecology

Question	Marks	Reference
Define *biomass.*	1	P3 HL M09 q2a P3 SL TZ1 q3a
Explain the biomass change in different trophic levels.	3	P3 SL TZ1 M09 q3c
Outline the temperature and vegetation characteristics of **one** major biome.	2	P3 SL TZ1 M09 q2b
Outline the major differences in temperature and moisture that are characteristic of two **named** biomes.	2	P3 HL TZ2 M09 q2a
Outline the characteristics of a tropical rainforest biome.	3	P3 HL N09 q2b
State the units used in a pyramid of energy.	1	P3 SL N09 q2b
Explain the small biomass of organisms in higher trophic levels.	2	P3 SL N09 q2c
(i) State **one** example of a deliberate release of an alien species, including the name of the organism and where it was released.	1	P3 SL TZ2 M09 q3ai
(ii) Using the example from (i) above, outline the reason for its release and the impact it had on the environment.	2	P3 SL TZ2 M09 q3aii
(i) Define *biomagnification*.	1	P3 HL N09 q2ai
(ii) Outline a **named** example of biomagnification.	2	P3 HL N09 q2aii
Discuss the impact of an alien species on an ecosystem.	3	P3 SL TZ1 M09 q2a
Discuss the impacts of a **named** alien species introduced as a biological control measure.	3	P3 SL N09 q3a
Outline the effects of ultraviolet radiation on living tissues.	2	P3 SL N09 q3b